MATERIAL REVOLUTION II

SASCHA PETERS

MATERIAL REVOLUTION II
NEW SUSTAINABLE AND MULTI-PURPOSE MATERIALS FOR DESIGN AND ARCHITECTURE

BIRKHÄUSER
BASEL

CONTENTS

I
INTRODUCTION

II
MATERIALS

III
APPENDIX

1 BIOPLASTICS AND BIO-BASED BONDING AGENTS

2 NATURAL MATERIALS AND ORGANIC WASTE MATERIALS

3 RECYCLING MATERIALS

4 LIGHTWEIGHT CONSTRUCTION MATERIALS

5 MULTIFUNCTIONAL MATERIALS

6 MATERIALS THAT INFLUENCE AND EMIT LIGHT

7 ENERGY-GENERATING MATERIALS AND INNOVATIVE INSULANTS

8 INNOVATIVE AND SUSTAINABLE PRODUCTION PROCESSES

6

THE FUTURE OF SUSTAINABLE PRODUCT DEVELOPMENT

It seems that our product culture built on the lavish overuse of the world's resources is outdated. The global population has now reached nearly 7.2 billion. This, along with rising consumer demand from developing countries, India and China, means it is no longer possible to ignore the facts: the world's resources are finite, and running out. The effects of this on the sourcing of raw materials for production were first felt during the economic recovery following the end of the crisis in 2010. Manufacturing materials such as high-performance plastics for vehicle construction became rare, precious, and almost impossible to come by. Comparable, even, to the rare earths used by the renewable energy sector to carry out advanced energy-conversion processes, producing the energy turn of which we are so in need.

Scientists have predicted that if we maintain our product culture and our current rates of consumption, as soon as 2030 we would need the equivalent of two planet Earths to meet our needs. As a result, a growing importance is being attached to recycling resources. The term "urban mining" is now being used to refer to the way that heavily populated cities can be mined for energy. Alongside this, an ever increasing value is being placed on the recyclability of products and materials. After analyses of toxins released by wood-based products in our apartments and plasticizers in polymeric materials, as well as the harmful effects of plastic waste in the ocean, concerned consumers are ever more inclined to seek out naturally sourced products. "Sustainability" is no longer merely a selling point. It is a new, real pressure to which our industrial culture is bound to respond.

The trend towards sustainable product development and sustainable design is especially important in the design industry, where it has been embraced by industrial designers and architects. More and more, creative industries are meeting the needs of a sustainable product culture by incorporating the latest scientific discoveries into their work. This brings research, technology, and design together, particularly within the allocation of materials.

NATURALLY OCCURRING AND BIODEGRADABLE

Designers are currently forming a new understanding of how products might be sourced and produced. They are looking increasingly towards producing in line with nature, and have made biological biodegradability and natural recyclability top priorities when seeking out new materials.

Grown fiber structures with black strawberry (Source: Carole Collet)

British designer Carole Collet sees the future of textile jewelry in the use of roots and shoots for the design of textile-based jewelry pieces and is currently testing the potential of plant growth in design. Examples of her creations are works based on black strawberry plant roots or the red shoots of tomato plants.

Bio-light (Source: Philips Design) → p. 167

At the end of 2011, Philips designers Clive van Heerden and Jack Mama investigated the potential use of bioluminescence for household lighting, a concept they presented as "bio-light" at the Dutch Design Week. Hand-blown glass structures drew up a liquid containing luminescent bacteria. The nutrients essential for the pale-green bioluminescence were supplied via silicone tubes and came from compostable waste directly from the kitchen.

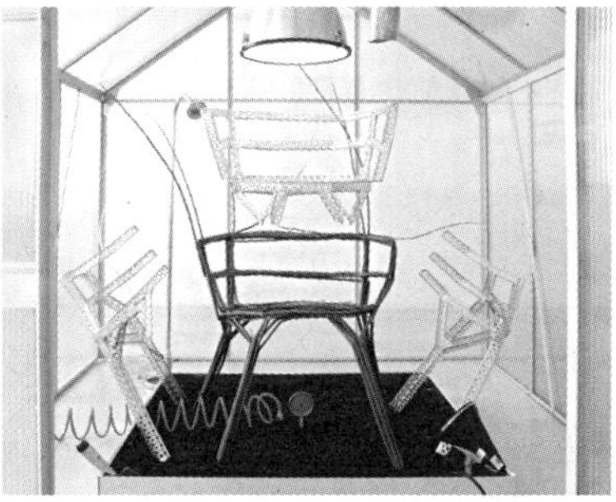

Chair Farm (Design: Werner Aisslinger)

Since the concept of "urban farming" has proved to be increasingly popular throughout the world, Berlin designer Werner Aisslinger has introduced a revolutionary production principle for the future of furniture design. At his Chair Farm, he uses perforated sheet steel as the frame for chairs made of bamboo shoots. The plants then grow in the direction which he has set up for them. With this, he defies the force of gravity, using waste materials to change the natural shape of the plant to suit his needs.

Stool "Xylinum" with bacterial cellulose coating (Design: Jannis Hülsen)
→ p. 044

The stool "Xylinum" is a piece of furniture by industrial designer Jannis Hülsen, for which he worked with Jenpolymers Ltd. to develop a 100% biodegradable coating system for wooden components. They used the bacteria culture xylinum, in which cellulose fibers grow when placed in a solution a purely biological process. The fibers link up into a tight thread-like structure and, once dried, form a leathery coating which looks plant based and natural.

Stool with biomass consisting of mycelium (Source: Phil Ross)

In 2012, artist Phil Ross from San Francisco optimized the quality of fungus-based materials to the extent that he was able to use them to produce seats and blocks for architectural structures. In so doing he followed in the footsteps of New York-based company Ecovative Design, which in recent years has developed a foam material from organic waste and mycelium.

Cookie Cup (Design: Enrique Luis Sardi)
→ p. 084

Cookie Cup is an edible espresso cup by designer Enrique Luis Sardi, developed together with pâtissier Cataldo Parisi. The design is part of a trend towards reducing the use of resources for serving food, both by having less packaging and containers and by keeping products biodegradable where possible. The edible cup is made from shortcrust pastry, meaning that the coffee flavor infuses into it, heightening the overall enjoyment of the coffee. An insulating layer of sugar glaze on the interior of the cup makes it temporarily watertight and so fit for use.

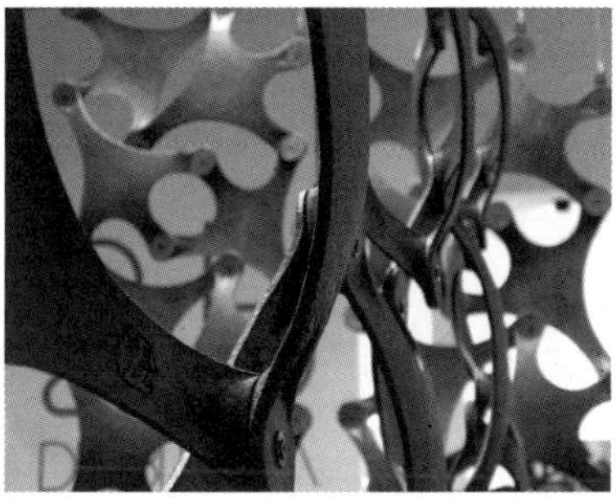

Structure made of FluidSolids® (Source: Beat Karrer)

In his material FluidSolids, architect Beat Karrer from Zurich uses a protein-based binding agent to convert natural fiber materials in industrial waste into a material which can be processed to make exhibition stands and furniture. The high molding accuracy of this material means that it can be used for classic molding techniques such as injections and extrusions. The material basis and the processing techniques are emission free, making the developers able clearly to quantify their low energy use in comparison with conventional products.

USING RECYCLED MATERIALS

Lamps made from coffee grounds, furniture from paper pulp, or flip flops from palm tree bark: natural waste materials are currently proving very popular in product and furniture design. The growing hope to see a clean and ecologically sound world now seems to have taken flight, leaving the supermarket shelves and entering the creative industries. As well as using biological waste, designers and architects are currently working on developing panel materials, hoping to create a fully biodegradable wood substitute.

EcoSystem natural fiberboard made of 100% renewable raw materials (Design: UdK Berlin) → p. 108

One example is by designers at the Berlin University of the Arts (UdK Berlin), who have created a natural fiberboard, EcoSystem. It is based entirely on renewable materials from agricultural waste and, unlike conventional wood-based materials, uses a bioplastics-based binding agent. The fiberboard is recyclable, biodegradable, uncoated, and without any of the harmful adhesives, varnish, or covering products that might be expected.

Biodegradable “Animal Coffin” (Design: Louise Knoppert, Flore de Maillard, Amanda Österlin La Mont, Christian Frank Müller)

The “Animal Coffin” is a biodegradable coffin for an animal, made from a mix of natural waste products and substances. The mixture of starch, flour, cardboard, coffee grounds, vinegar, and hair guarantees the compostability of the structure, and makes it strong enough for use. Moreover, the lid of the coffin has been planted with seeds which will grow into healthy trees or plants once it has decayed. “Animal Coffin” was presented with an excellence award at the 2012 Adream Competition.

Biodegradable Compos Chair (Design: Samuli Naamanka)

Compos Chair, by Finnish designer Samuli Naamanka, was also made without the use of a petrochemical binding agent. Naamanka has created a seat shell using a natural fiber composite material, which is 100% biodegradable and emission free. Here, cornstarch is used as the binding agent, which is polymerized during the production process.

Decafé luminaires made of coffee grounds (Design: Raúl Laurí) → p. 078

In spring 2012, designer Raúl Laurí impressed experts in the field with his design for lampshades made from coffee grounds, for which he was awarded the first prize for up-and-coming designers at the Milan Furniture Fair. A natural binding agent holds the organic materials together and makes the design biodegradable. As well as lamps, the designer has also produced a table and crockery set from the waste material.

Palmleather made of areca palm leaves (Source: Tjeerd Veenhoven)
→ p. 084

Palmleather is the name that designer Tjeerd Veenhoven has chosen for a material that he has extracted from areca palm leaves and is using to make bags, sandals, and book covers. The leaves are soaked in a biological solution and their natural plant oils are then released. When left for a long time, the fibers become soft and flexible. The designer commissions his designs to be manufactured in southern India.

LIGHTWEIGHT AND RESOURCE FRIENDLY

"Lightweight construction" has become a key phrase in today's sustainability debate in terms of reducing our resource consumption. Through the consistent use of innovative multimaterial concepts, substantial progress has been made in terms of creating lighter and more efficient products, leading to savings on resources, and therefore energy, in both production and transport. Designers and architects are currently focusing on developing solutions and structures able to incorporate materials into lightweight construction in ways we would never have thought possible. Moreover they are showing how lightweight construction can advance electromobility concepts.

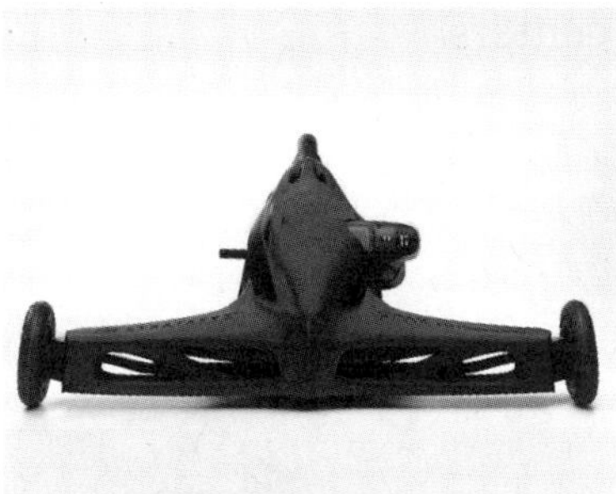

Printed lightweight construction powered by a cordless screwdriver (Design: HAWK Hildesheim)

For the 2011 Cordless Screwdriver Competition, the Design faculty at the HAWK University of Applied Sciences and Arts Hildesheim created a vehicle using a 3D printer. The vehicle was built from a single piece of acrylonitrile butadiene styrene (ABS) in the space of 10 days, in layers just 0.25 mm thick. By working with bionic internal structures, the designers were able to use material only in those places where it was absolutely necessary to ensure stability. This enabled the vehicle to have a total weight of only 6 kilograms.

Carbon-fiber reinforced concrete furniture (Source: Paulsberg)
→ p. 115

Lars Schmieder is one of the first designers to examine the potential of high-strength carbon fibers in concrete for furniture design. At his production facility in Dresden, he created ultrathin concrete armchairs and tables using carbon-fiber reinforced structures only a few millimeters thick. This enabled him to have more scope in his design; he could add indentations to the chairs, in elegant slants and bends.

Carbocrete Balconies (Design: Stefan Paulisch, Uta Kleffling, Pamela Voigt)
→ p. 115

In order to extend the applications of fiber concrete even further, in spring 2012 SGL Carbon held an open innovation competition. Designers and architects came up with around 300 ideas for the composite material Carbocrete. The winners of the competition were Stefan Paulisch, Uta Kleffling, and Pamela Voigt, a design team from Leipzig, with their "Carbocrete Balconies." These organically curving balconies, which can be used for planting, were a way for the design team to bring new life to the concrete wastelands we so often find in city centers.

Lightweight building component (Source: Jens-Hagen Wüstefeld)

Architect Jens-Hagen Wüstefeld's spatial lattice structure consisting of interconnecting triangles absorbs forces from all directions and distributes them across the bordering surfaces and edges, thus ensuring optimal force distribution. The structure enables a weight reduction of 85% compared with solid material. Round, spherical, and contoured components can be created in this way, in all kinds of materials. The simple process involves making slanted incisions in strips of the chosen material, which are then interlocked with one another.

Sandwich construction consisting of pieces of bamboo cut at an angle (Design: Wassilij Grod)

The panel material by designer Wassilij Grod is a sandwich construction with a central layer of diagonally cut bamboo tubes, glued onto a top and bottom layer. This structure is highly pressure resistant while using fewer resources. The blend of slanted bamboo pieces keeps resource needs to a minimum. The strength of the panels can be influenced by variations in the arrangement of the rings.

SMART, DYNAMIC, ENTERPRISING

It is becoming increasingly interesting for designers to incorporate functions into materials or composites. Until recently, the use of so-called smart materials had been very restricted, but the latest advances in creative freedom have given designers and architects free rein. Creative designs with varying qualities that can be easily influenced are thus becoming just as ecologically important as products that generate their own energy or that themselves serve to improve environmental conditions. Furthermore, creating products with smart materials has allowed designers to experiment and discover new manufacturing methods.

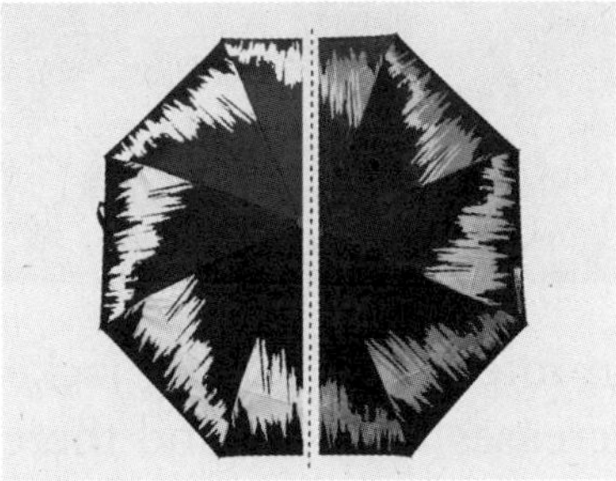

Water-sensitive umbrella (Design: SquidLondon)
→ p. 127

Using water-sensitive pigments, the designers at SquidLondon have developed a raincoat that changes color when it gets wet. In this way the creative professionals show how smart materials can react, for instance, to the weather.

Sugru modeling clay (Source: Jane Ní Dhulchaointigh)

Sugru is a self-hardening silicone modeling clay by Irish product designer Jane Ní Dhulchaointigh. The material was voted one of the 50 best innovations by TIME Magazine in 2010 and offers, particularly for designers, real potential to be a universal modeling material. Sugru can be modeled over 30 minutes, after which time it hardens in the air and takes on the consistency of hard, slightly elastic rubber. The silicone material is self-adhesive on almost all material surfaces, forms strong joints, and can be used as a glue and an insulation material. The clay is nontoxic and after it hardens it is water-resistant and heat resistant up to 180 degrees Celsius.

Functional Food Fictions (Source: Helge Fischer, Ann-Kristina Simon)

The designers from Bold Futures have specialized in developing future scenarios for new technologies. One example of Helge Fischer and Ann-Kristina Simon's creations is a fruit gum with nanoparticle-encapsulated substances capable of sobering up eaters in the space of just 20 minutes. Other products include baked goods designed to simulate symptoms of disease, and yoghurt products for pregnant women which are claimed to be able to encourage certain traits in their unborn child.

Interactive cycling jacket "Sporty Supaheroe" (Design: Wolfgang Langeder) → p. 133

Sporty Supaheroe is an interactive cycling jacket with integrated LED illumination, which increases visibility and therefore also safety on the road. Designer Wolfgang Langeder worked with Stretchable Circuits and the Fraunhofer Institute for Reliability and Microintegration (IZM) to integrate a flexible display into the clothing, meaning that up to 64 RGB LEDs can be incorporated. The lights can be directed by the cyclist, as a result of sensors that respond to body movements before transmitting control signals to the lights.

VIVID light installation consisting of white nylon airbags (Source: Julia Berner, Alexander Dronka, Johannes Roloff)

At the trade fair "Light and Building 2012" in Frankfurt, students from HAWK Hildesheim envisaged a remarkable concept study for the future of light, namely an interactive cloud of white nylon airbags, that hovers over the observer. In their original state, all of the airbags look alike. But, equipped with fans, lamps, and sensors, they are able to respond to people which triggers different reactions in them. If one of them is touched, it will inflate and start to glow.

ADDITIVE GENERATION

We have been aware of the potential of generative manufacturing since the end of the 1980s. In the context of current attempts to keep resource use to a minimum in building components and architectural structures, additive technologies are playing an ever greater role. Experts calculate that compared with classic production methods, these technologies can reduce the weight of products from between 50 and 90%. In particular, as they can be used to realize highly complex shapes very simply, generative technologies are gaining ever more popularity in creative industries. It is not only through so-called rapid technologies that designers and architects are making changes, however; they are developing new principles, allowing materials to transform the design process itself.

Gravity Stool (Design: Jólan van der Wiel)

The extraordinary shape of the Gravity Stool is the result of exploiting gravity in a magnetic field. To make it, designer Jólan van der Wiel set up a production facility with a huge magnet, which he presented at the IMM 2012 in Cologne. Van der Wiel initially mixes a plastic compound with a magnetic powder. Then the combination of the forces in the magnetic field and of gravity produces shapes that until now we had only seen in nature.

Organoid lounge furniture IOYO (Design: Nofrontiere)

In collaboration with developers from Organoid Technologies, the designers at Nofrontiere in Vienna have created seating from finely ground natural residues such as wood shavings, grasses, or nutshells. Together with an organic binding agent, this natural waste is then injected into a negative mold and hardened in a vacuum. The resulting three-dimensional molds are fully compostable.

The metabolic factory (Design: Thomas Vailly)

For Milan Design Week 2012, Thomas Vailly worked under the name of "The metabolic factory," presenting a procedure which allowed cups to be made out of human hair. The hair was mixed with glycerin and sodium sulfate, forming a natural material similar to leather. This was then formed into different shapes, which, of course, are compostable.

Stone Spray structures made of sand (Source: Anna Kulik, Petr Novikov, Inder Shergill)
→ p. 197

Anna Kulik, Petr Novikov, and Inder Shergill, students at the Institute for Advanced Architecture of Catalonia in Barcelona, have been working on a project called "Stone Spray" since 2012, designing a robot that is in future intended to realize highly resilient structures based on additive construction techniques using sand and a binding agent from the field of road building. Tests using a base fabric to apply the sand mixture have shown promising results. The shape of a stool, for example, could be created in as little as about three hours. Given that the "sand sprayer" can be used on site, it could revolutionize building techniques.

Printed space station on the Moon (Source: Foster + Partners)
→ p. 197

Developers working for renowned London-based architect Norman Foster and the European Space Agency (ESA) are currently testing 3D printers using lunar materials, such as the lunar mineral regolith, to make architectural structures on the Moon. To simulate the conditions, a dome structure was designed from walls with cellular structures, sprayed in layers by means of the print nozzles on a 6 m high frame with sand-like particles and a strong binding agent.

Carbon-fiber reinforced concrete furniture
(Source: Paulsberg)
Introduction // Lightweight Construction Materials
→ p. 011

Lampshades made of coffee grounds
(Design: Raúl Laurí)
Coffee Ground Materials // Natural Materials and Organic Waste Materials
→ p. 079

Chair Farm (Design: Werner Aisslinger)
Introduction // Natural Materials and Organic Waste Materials
→ p. 007

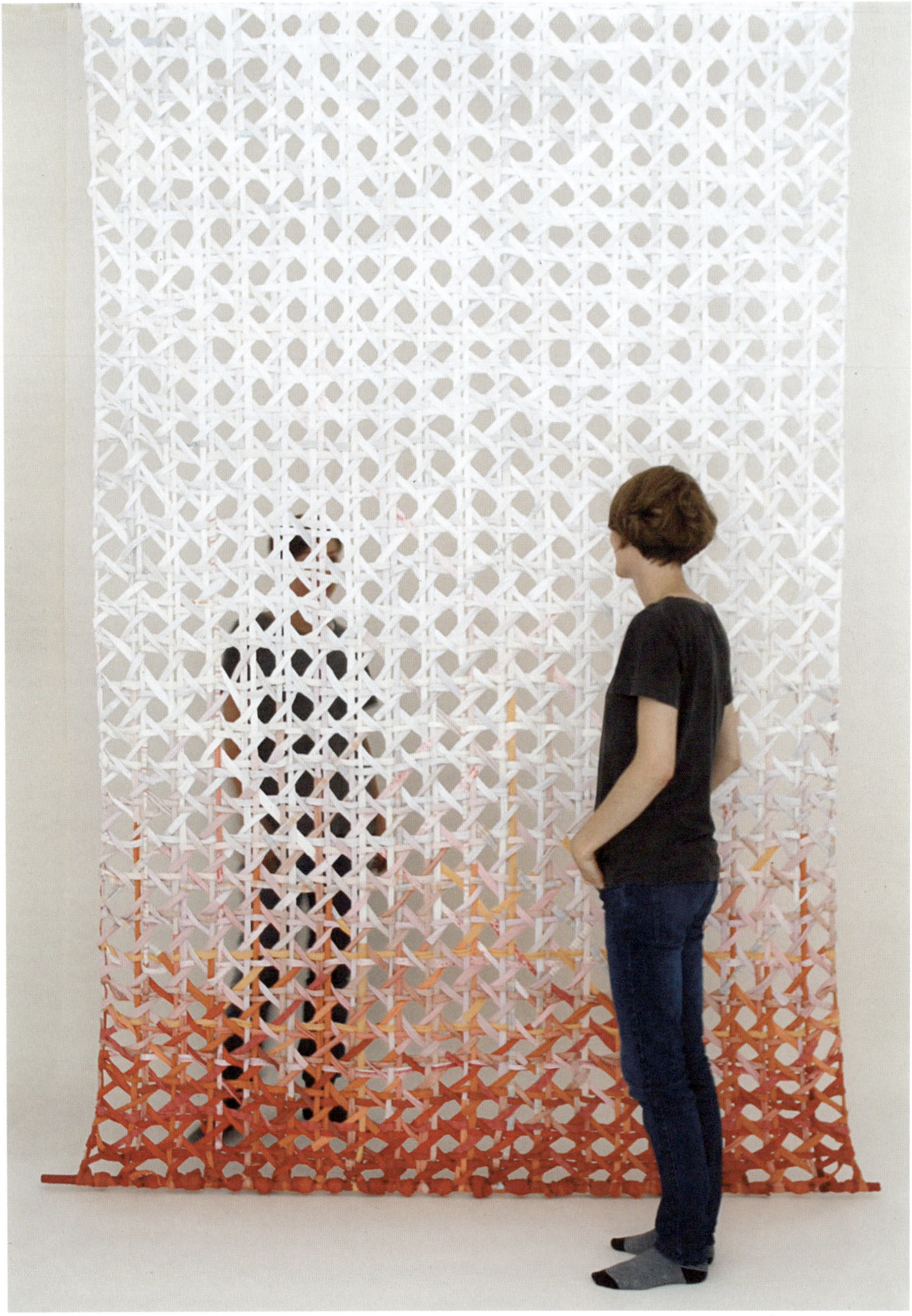

REMÖTIL — Fabric wall made of recycled textiles
(Design: Moa Hallgren and Lisa Spengler)
Waste Textile Materials // Recycling Materials
→ p. 095

Packaging made of recycled paper fibers
(Source: Flextrus)
Waste Textile Materials // Recycling Materials
→ S. 095

VEIO textile bowls
(Design: Kathrin Morawietz)
Waste Textile Materials // Recycling Materials
→ p. 095

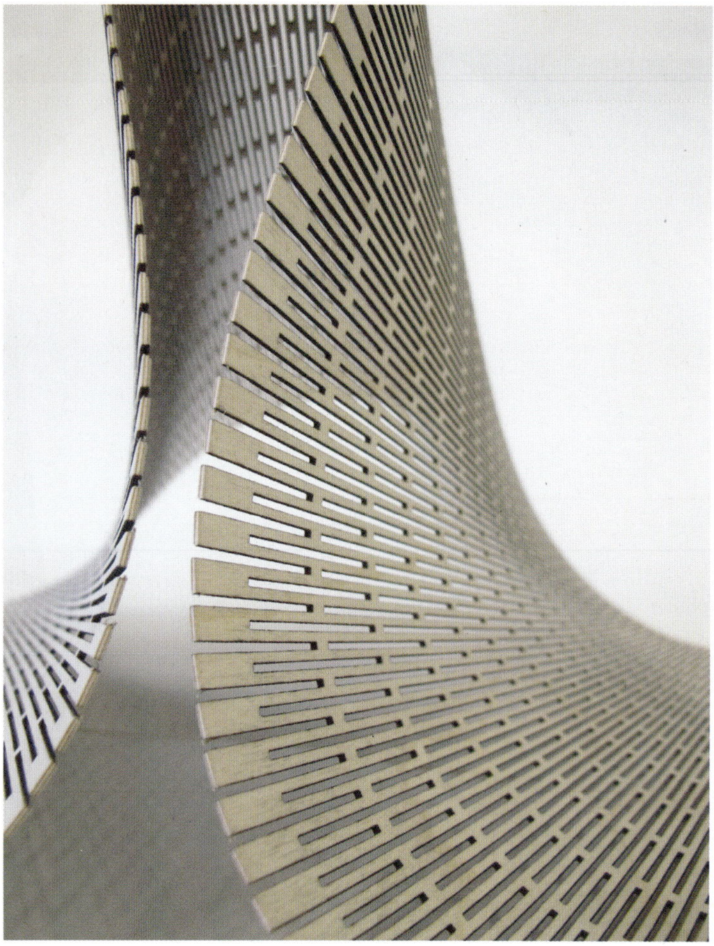

Incisions can reduce the weight of woodbased materials (Source: Dukta)
Weight-optimized Timber Materials and Replacement Materials // Lightweight Construction Materials
→ p. 108

Flexible perforated timber (Source: Dukta)
Weight-optimized Structured and Honeycomb Constructions // Lightweight Construction Materials
→ p. 111

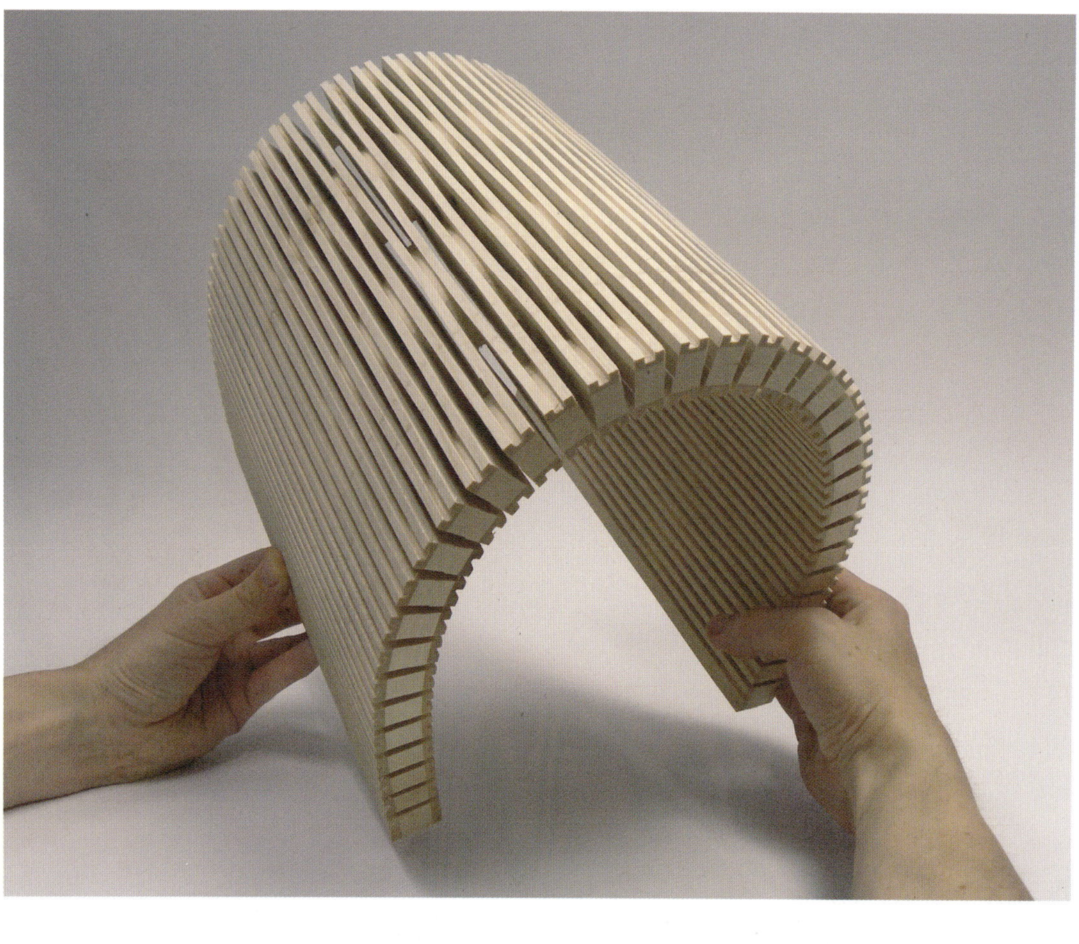

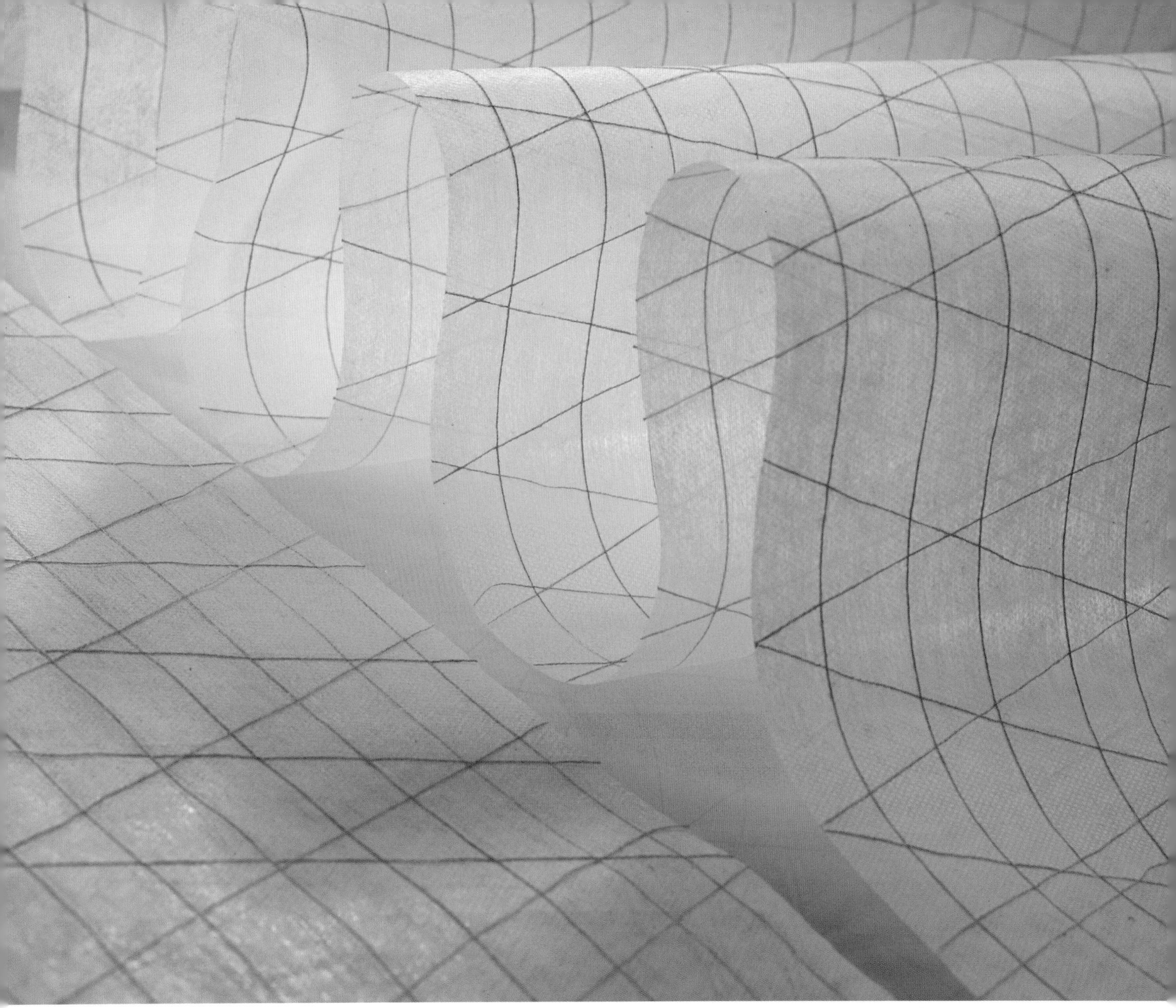

BafaTex filter non-woven material (Source: BafaTex)
Folded Lightweight Construction // Lightweight Construction
Materials
→ p. 113

Multistory car park in Montreux with Tensairity pneumatic structure
(Source: Empa, Architecture: Luscher Architectes SA & Airlight Ltd.)
Biomimetic Lightweight Construction // Lightweight Construction Materials
→ p. 119

Air dome
(Source: Paranet Germany)
Pneumatic Textiles // Lightweight
Construction Materials
→ p. 120

Airdrop — irrigation system based on the Namib Desert beetle (Design: Edward Linacre)
Water-collecting Surfaces // Multifunctional Materials
→ p. 141

Screen "And A And Be And Not" made of dichroic glass
(Design: Camilla Richter)
Light-directing Materials // Materials that Influence and Emit Light
→ p. 159

Mobile light installation using laser-cut EL films and based on man-machine interaction (Source: CAAD, Manuel Kretzer)
Electroluminescent Materials // Materials that Influence and Emit Light
→ p. 164

Bio-Light with luminescent bacteria that have been fed on methane and organic compost (Source: Philips Design)
Biological Light // Materials that Influence and Emit Light
→ p. 167

VIVID light installation consisting of white nylon airbags
(Source: Julia Berner, Alexander Dronka, Johannes Roloff)
Introduction
→ p. 014

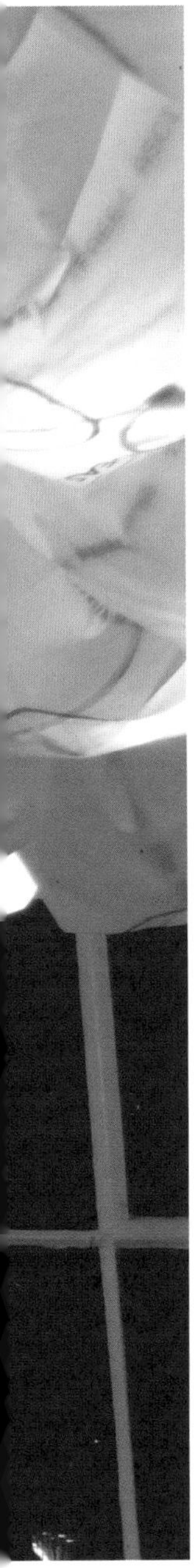

Endless Flow — Generative furniture production with recycled plastics
(Design: Dirk Vander Kooij)
Innovative and Sustainable Production Processes // Generative Manufacturing with Recycled Materials
→ p. 191

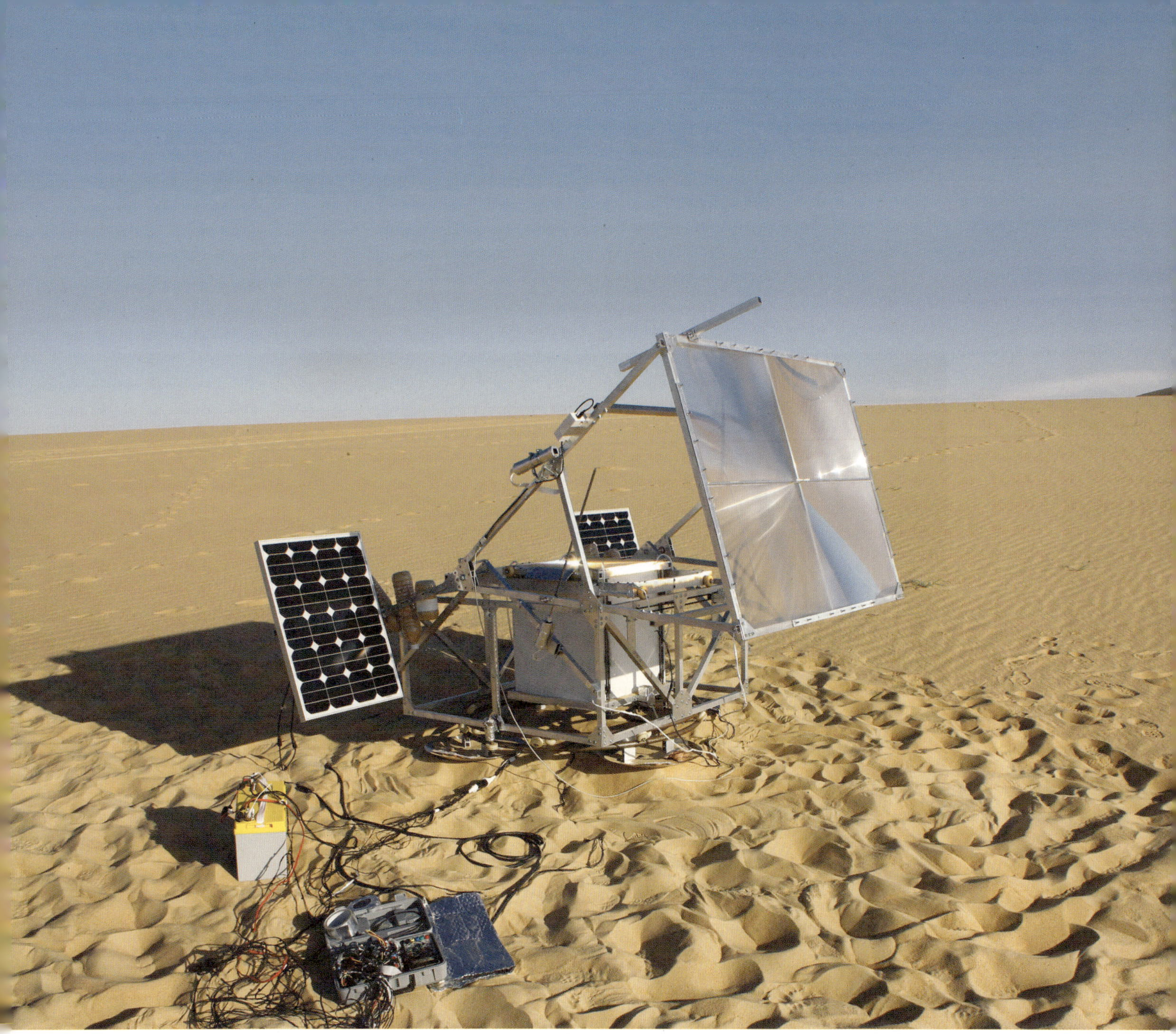

Solar Sinter in the Egyptian desert
(Source: Markus Kayser,
photo (above right): Wendelin Schulz-Pruss)
Solar Sinter Rapid Manufacturing
through Sunlight // Innovative and Sustainable
Production Processes
→ p. 190

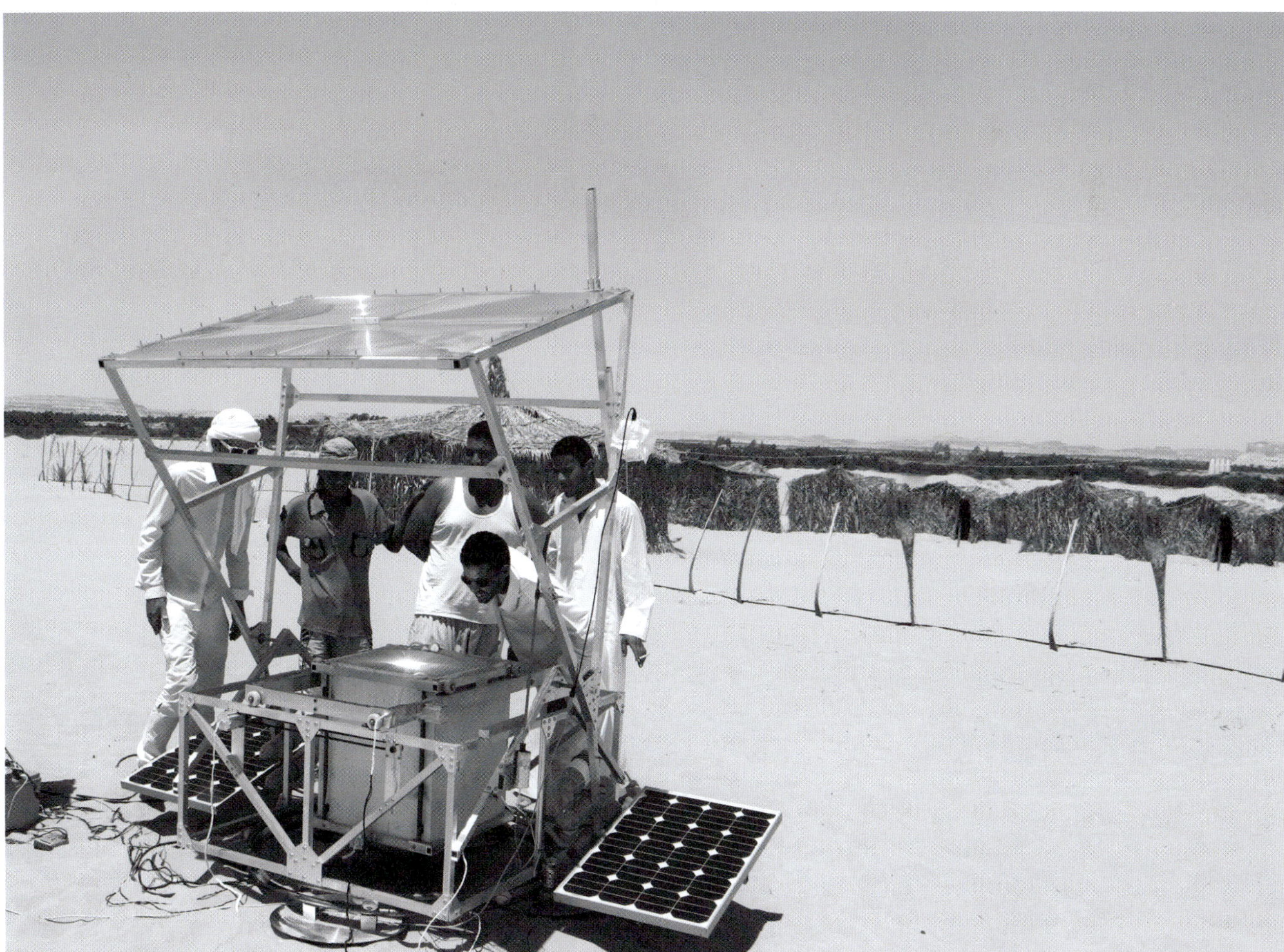

Gravity Stool
(Design: Jólan van der Wiel)
Introduction
→ p. 015

BIOPLASTICS AND BIO-BASED BONDING AGENTS

01

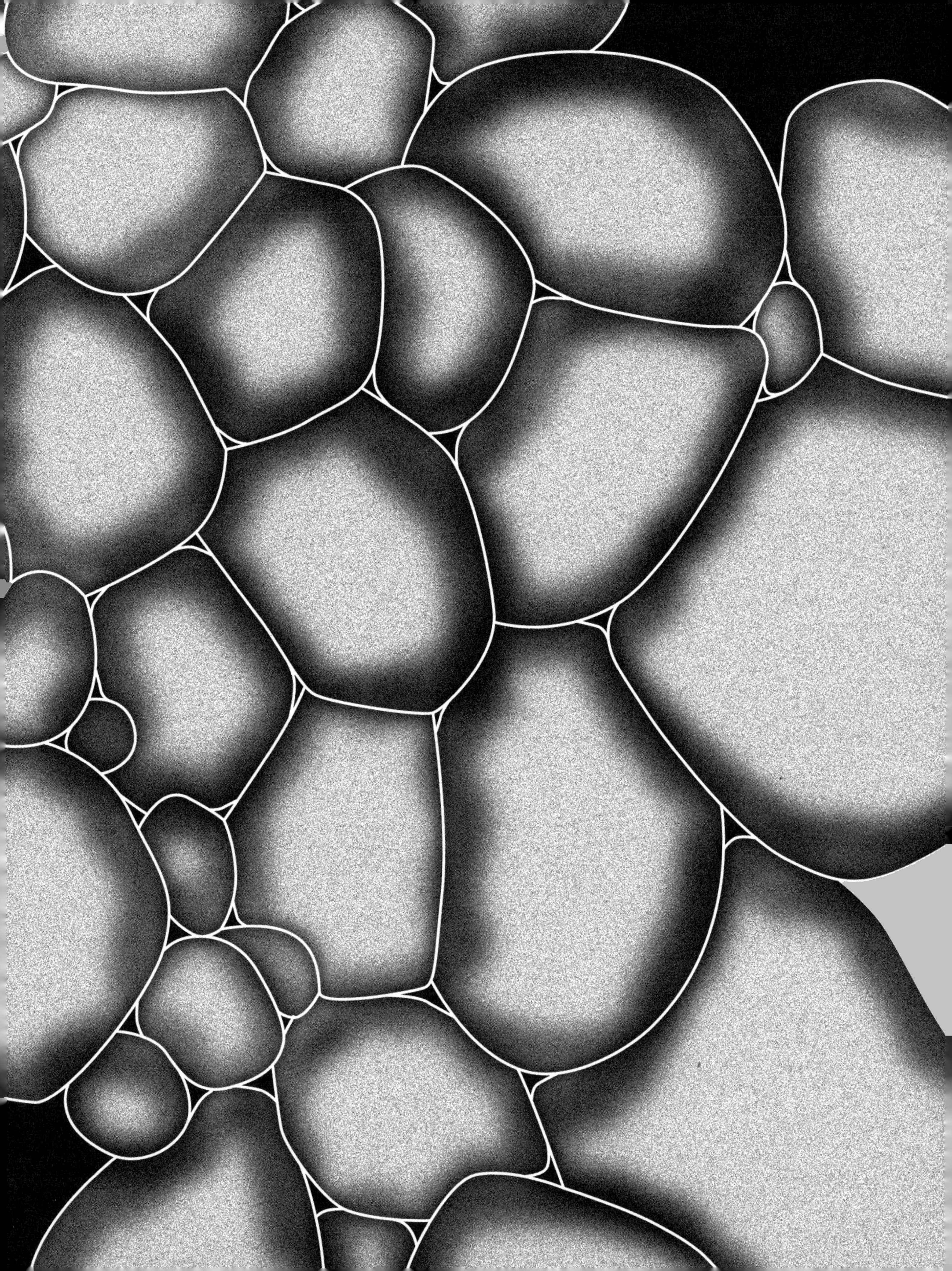

Increasingly frequently we hear that oil resources are becoming scarcer. Although new deposits are always being found, the tapping of which is now profitable due to rising prices (for example, oil drilling in Mecklenburg-Vorpommern), the eventual end of the petrochemical industry is inevitable. Chemical corporations are already working intensively to prepare for a paradigm shift, which will result in an abandonment of fossil raw materials for our product culture in favor of bio-based production methods. While the question of what the future **bio-economy** might look like is still a subject of political debate, producers are building up capacities for the manufacture of bio-based products.

A current study by the Institute for Energy and Environmental Research in Heidelberg, carried out on behalf of the Federal Environment Agency, presents a positive outlook for the bioplastics industry: the environmental impact profiles of a number of bioplastics have improved significantly according to scientists, and further potential for optimization is currently being exploited. When it comes to an ecological audit, bioplastics often consume less fossil fuel resources and produce less CO_2 emissions compared with their conventional counterparts.

The confidence of the markets in the bioplastics industry is evident from the most recently published market data (Source: European Bioplastics, October 2012). The data shows that global production capacities for bioplastics are likely to increase fivefold between 2011 and 2016, from around 1.2 million tons in the year 2011 to almost 6 million tons in 2016. In the field of packaging, in the long term up to 70% of conventional plastics are likely to be replaced by bio-based alternatives.

37

BIOPLASTICS AND BIO-BASED BONDING AGENTS

POTENTIAL AND PRODUCTION PROCESSES

The greatest potential to be found among biodegradable plastics, manufacturers believe, is predominantly among those based on polylactic acid (PLA) or polyhydroxy fatty acids (PHF). This is because they have the qualities to replace classic bulk plastics like polyvinylchloride (PVC), polyethylene (PE), or polypropylene (PP) in the medium term. The market now offers a broad range of bio-based plastics that stem from various production processes and base materials.

In the Brockhaus encyclopedia, bioplastics are classified as "plastic-like materials" which can be produced "entirely or almost entirely from biopolymers and worked using conventional plastics processing methods." However, there is a widespread understanding that bioplastics also include those that are not based on renewable sources but which can be broken down biologically into natural substances such as oxygen, water, or compost. The use of bioplastics offers a way for the industry to save fossil resources, use them more efficiently, and reduce the carbon footprint of plastic products.

Manufacturers have pursued various strategies in the production of bioplastics. In order to be able to use conventional production processes on a large scale (drop-in solutions), a number of chemical companies are making use of base chemicals (platform chemicals) that can be produced using renewable raw materials. This applies particularly to ethanol, which is used as a basis for thermoplastics such as PE, PP and PVC. Bioethanol can now be produced from sugarcane or sugar beet. Commercial-scale production of bio-PE is underway, predominantly down to Braskem, a Brazilian petrochemicals company and global giant in polymer production. Production of bio-PA, bio-PP, and bio-PUR is being built up and will be expanded over the next few years. The greatest share of the market, some 40%, is held by partly bio-based polyethylene terephthalate (PET). Its huge significance and a tenfold increase in production capacity will increase this share to a predicted 80% by 2016 (Source: European Bioplastics, October 2012).

Material coordinate system for bioplastics
(Source: Hanover University of Applied Sciences and Arts, Hans-Josef Endres)

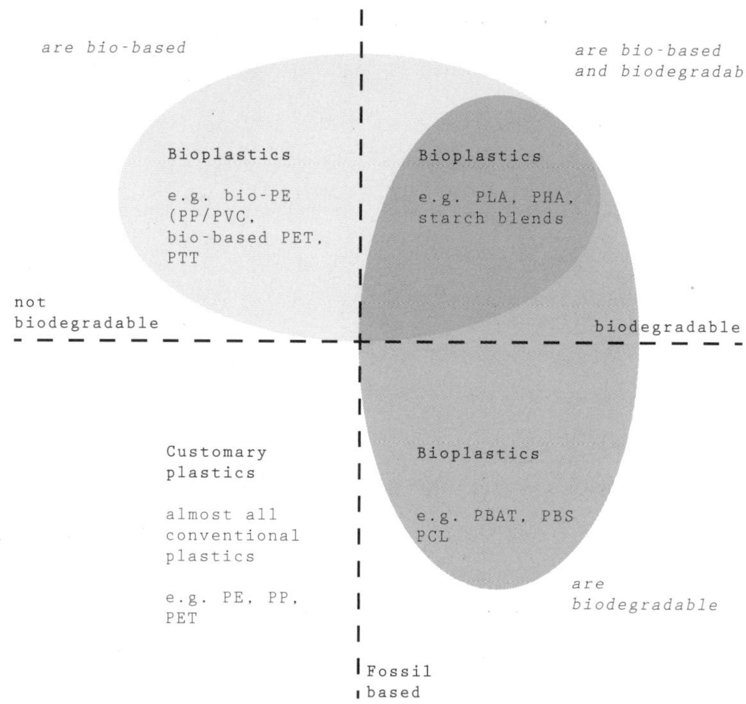

Global production capacities for bioplastics, 2012
(Source: European Bioplastics)

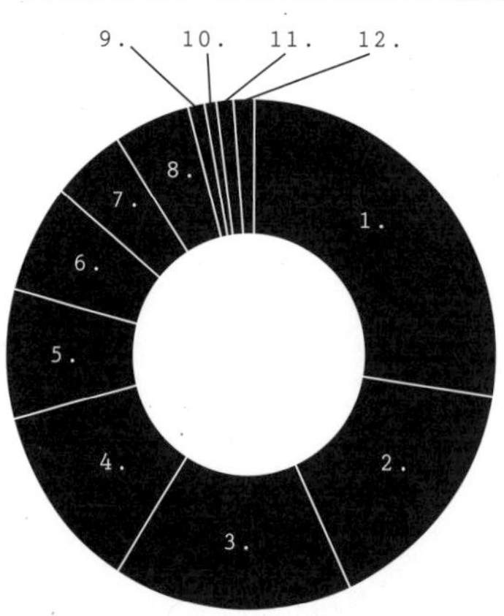

Global production output for bioplastics 2012 by type of material (Source: European Bioplastics)

1.Bio-PE	200,000	28%
2.Biodegradable starch blends	117,800	16%
3.PLA	112,500	15%
4.PHA	88,100	12%
5.Biodegradable polyester	56,500	8%
6.Bio-PET	50,000	7%
7.Cellulose regenerates	36,000	5%
8.Bio-PA	35,000	5%
9.Cellulose derivatives	8,000	1%
10.PLA blends	8,000	1%
11.Durable starch blends	5,100	1%
12.Others	7,500	1%
Total	724,000	100%

In addition to the chemical synthesis of natural base materials, manufacturers can also produce bioplastics through direct synthesis of biopolymers or through the modification of renewable raw materials. Among the bioplastics synthesized through direct fermentation, PHF and polyhydroxyalkanoate (PHA) appear particularly promising for potential applications. Bioplastics produced through the direct modification of renewable raw materials include those based on cellulose, starch, lignin, vegetable oils, chitin, and animal proteins. Most recently, there has been some success in the development of protein-based production processes. Most worthy of mention in this regard are milk protein fibers with their antibacterial effect, or the production of extremely tear-resistant and, at the same time, extremely ductile fibers based on spider silk protein.

"Grapple" clothes hook containing a high proportion of grass fiber (Design: Ryan Frank, source of material: Agriplast)

Bioplastics based on gelatin (Source: CAAD, Manuel Kretzer)

Properties bio-PET based on sugarcane molasses // lightweight // no plasticizers // prevention of evaporation due to silicon dioxide coating

Sustainability aspects better ecological balance of PET plastic bottles // based on renewable raw materials // possibility of water sterilization

BIO-BASED POLYETHYLENE TEREPHTHALATE (BIO-PET)

PET is one of the bulk plastics used in packaging. Most of us are familiar with the material from its use in plastic bottle production. Compared with glass bottles, PET bottles are lightweight, less easily broken in transportation, and can be recycled when carefully separated from other waste. Since 2010, bottles have also been made using partly bio-based PET.

MATERIAL CONCEPT AND PROPERTIES

In the form of a thermoplastic polyester, PET is produced from polycondensation of the monomers **terephthalic acid** and **ethylene glycol.** When it comes to the partly bio-based variants, ethylene glycol is made using **sugarcane molasses,** while the **terephthalic acid** is still produced by a petrochemical process for cost reasons. Recently, manufacturers announced the development of direct alternatives to PET. These include, for example, polyethylene **furanoate** (**PEF**), which has been a target of investment in the Netherlands since 2011. PEF is superior to PET in a number of ways, demonstrating lower gas permeability and greater heat resistance.

PET bottles contain no potentially harmful **plasticizers** like **phthalates** or harmful **bisphenol A.** However, the odorous **acetaldehyde** (**ethanal**) does leak out of the plastic, which releases it into the liquid, although the Federal Institute for Risk Assessment rates this exposure as harmless. In order to seal up the PET bottles against gas permeation and the release of acetaldehyde, a **silicon dioxide coating** just a few nanometers thick has been developed for the inside of the bottles.

In the spring of 2012, drinks manufacturer Pepsi announced the development of a plastic bottle based on **switchgrass, pine bark,** and **corn** in a combination of biological and chemical processes, which is supposed to boast similar properties to PET manufactured by a petrochemical process. The new material is claimed to result in considerably reduced CO_2 emissions and can be completely recycled. Pepsi is currently testing other natural base materials, such as **oat husks,** and **potato** and **orange peel,** as bases for bottle production.

APPLICATION

PET multiuse bottles demonstrate a better ecological footprint across their full life cycle compared with similar glass containers. This is predominantly the result of their low weight for transportation, their energy-efficient production, and their recyclability. In developing countries, clear PET bottles are increasingly used for water sterilization. The **SODIS method** (solar water disinfection) is based on the germicidal effects of **ultraviolet radiation.** Other areas where PET is used are films and fibers.

Orange peel as a basis for bioplastics

Solar water disinfection in Indonesia (Source: Michigan Technological University)

Plastic bottles made of partly bio-based PET (Photo: Diana Drewes)

Properties identical properties compared to conventional material // low-cross-linked bio-PUR with bio-alcohol produced by fermentation // densely cross-linked bio-PUR with vegetable oil // biodegradability reduces with denser cross-linking

Sustainability aspects based on renewable materials // partly biodegradable

BIO-BASED POLYURETHANE (BIO-PUR)

Polyurethanes (PUR) are a very important group of plastics for various applications, which can have thermoplastic (thermoplastic polyurethane, TPU), duroplastic (polyurethane resins), and elastomer properties. As a consequence, some manufacturers believe it makes sense to produce PUR using renewable raw materials.

MATERIAL CONCEPT AND PROPERTIES

The conventional manufacturing process is based on a polyaddition of **alcohols** with **isocyanates.** Depending on the alcohol used, various degrees of cross-linking and thus the material's properties under the influence of heat and/or its elasticity can be determined. In the manufacture of **bio-PUR,** raw materials are used for the production of alcohol. For low-cross-linked PUR, a **bio-alcohol** produced through fermentation suffices. The basis of densely cross-linked bio-PUR is formed by **vegetable oils** such as **rapeseed, soya, castor,** or sunflower oil. Research is also being carried out on the development of the alcohol component using **lignin.** Bio-PUR boasts identical properties to its conventional counterparts. Increasing the **cross-linking level** reduces the possibility of partial degradability of the plastic.

APPLICATION

Possible uses for bio-PUR can be found wherever PUR, with its manifold properties, is used in large quantities. This applies to the foam materials sector in the production of mattresses or insulating materials, as well as in the shoe industry and the sports equipment sector. The complex geometries of soles, window profiles, or trim elements for motor vehicles are usually produced by means of **reaction injection molding** (RIM). During this process, the cross-linking of the PUR foam takes place only once the material is inside the mold.

Rape oil forms the basis of more strongly cross-linked bio-PUR (Photo: Diana Drewes)

Shoe production using a bio-based PUR (Source: Bayer MaterialScience)

BIO-BASED POLYAMIDES

Polyamides are among the most important plastics for technical applications. The growing scarcity of crude oil resources means the development of bio-based initial sources for manufacturing plastic is attracting increasing attention, which can lead to a smaller carbon footprint across the entire life cycle of the material.

Properties similar properties to polyamides produced petrochemically // entirely bio-based or largely so // not biodegradable

Sustainability aspects smaller carbon footprint // based on renewable raw materials

MATERIAL CONCEPT AND PROPERTIES

Examples available to date include the up to 100% bio-based **polyamide 10.10** (Vestamid Terra®), the completely bio-based **PA 4.10** (EcoPaXX®), and a **polyamide 6.10** (Ultramid® Balance or Zytel®), which is produced from up to around 60% renewable raw materials such as **castor oil.** They possess very good mechanical and physical properties, which can be improved further through the addition of fibers. In 2011, scientists at the Technical University of Braunschweig were able to optimize the use of the **soil bacteria** C. glutamicum to find an efficient method of **fermentational production** of **diaminopentane,** a key component of the 100% bio-based polyamide **PA 5.10**.

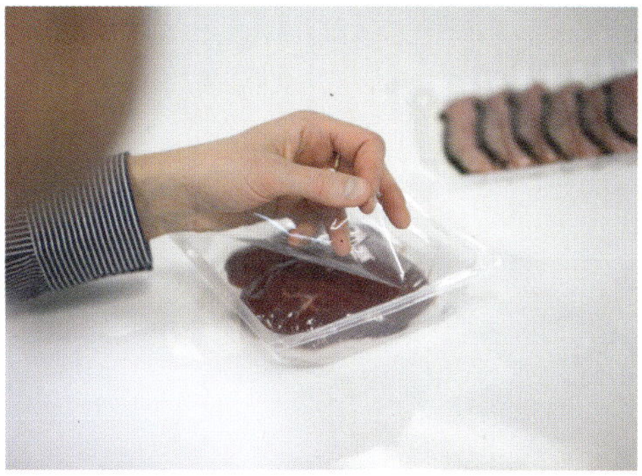

Packaging made of the bio-based PA 4.10 (Source: DSM Engineering Plastics)

Castor seeds (Source: DSM Engineering Plastics)

APPLICATION

Bio-based polyamides can generally be used in similar areas to their petrochemical counterparts. They are not biodegradable, so they are suitable for use in fuel and brake hosepipes for the automobile industry as well as for sports footwear, cable sheathing, and casing for anti-electrostatic appliances. Other typical uses for polyamides are in toothbrushes, hosiery, and medical technology. The market volume of bio-based polyamides is still something of a niche market, however.

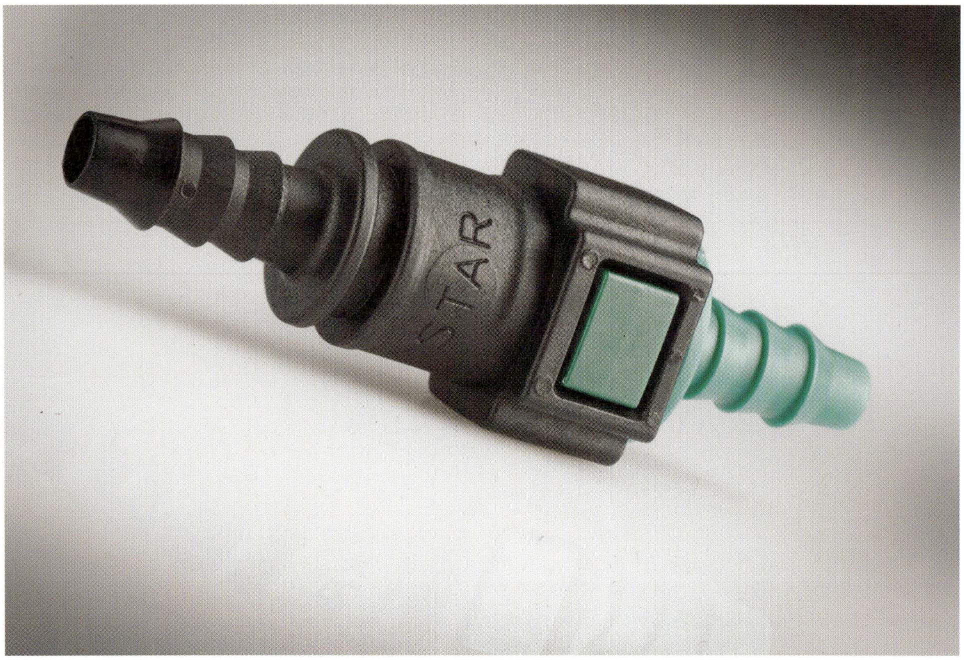

Quick-action coupling made of the fiberglass-reinforced polyamide 6.10 (Ultramid® Balance), which is not only hydrolytically stable, but is also resistant to fuels and zinc chloride (Source: BASF)

POLYHYDROXY FATTY ACIDS (PHF)

Properties similar properties to PVC, PE, and PP // thermoplastic processing // food-safe

Sustainability aspects biodegradable // based on animal waste

Alongside PLAs, it is predominantly PHFs (also known as PHAs) that show the greatest potential among biodegradable variations of bioplastics. As part of its ANIMPOL project, by the end of 2012 the Technical University of Graz had developed an industrial production process for their manufacture, based on fatty animal waste from the meat industry. This amounts to around 500 metric tons every year in Europe and has so far been thermally recycled in most cases.

MATERIAL CONCEPT AND PROPERTIES

During the production process, the elements that would negatively influence the quality of **biodiesel as a fuel,** i.e. those with **saturated fatty acids,** are separated and used as a raw material for the biotechnological fermentation of PHAs. PHFs are natural polyesters, which could replace the traditional bulk plastics such as PVC, PE, or PP. They are thermoplastic and can be processed in conventional systems used in the polymer industry (injection molding, extrusion, blow molding). Compared with PLAs, PHFs are more resistant to heat.

APPLICATION

The proportion of PHA in the bioplastics market amounts to around 5%, which is on the rise. PHFs are used, for example, to make biodegradable packaging, whose significance will increase given the reports on plastic waste in the sea. In medicine, too, PHA has already been used as a suture material that breaks down in the body.

PRODUCTS

Proganic®

This prizewinning thermoplastic biomaterial is largely made up of PHA, **carnauba wax,** and **mineral bulking agents.** It can be processed using conventional techniques and, by dint of being food-safe and water-resistant, is particularly suitable for consumer goods such as watering cans, disposable cutlery, flower pots, and egg cups. The material keeps its shape up to a temperature of 100 °C.

MINERV® PHA SC

This PHA is produced using waste products from the sugar industry (**molasses** or syrup). It is entirely biodegradable and, with its outstanding thermal properties, can be used as an alternative for the bulk plastics PET, PP, PS, or PE in the production of bottles, packaging, films, vehicle components, or electronics. Typical processing techniques include extrusion or injection molding.

Test set-up for the fermentation of PHAs based on animal fats (Source: Martin Koller, TU Graz)

PHAs based on animal fats (Source: Martin Koller, TU Graz)

BACTERIAL CELLULOSE

Properties finer than plant-based materials // no disruptive additional substances // biocompatible // no allergic reactions // highly tear-resistant, including when wet // significant shrinkage when drying // slow growth

Sustainability aspects production based on renewable raw materials // bacterial cellulose contains no harmful substances // biodegradable

Cellulose fibers are among the most important base materials of the textiles industry and come almost entirely from plant origins. With an alternative bio-based production process, microbes are used to convert glucose into cellulose within a fermentation process, producing a gel-like textile surface with thickness of up to 400 mm. A comprehensive range of bacteria strains is currently being researched which produce microscopically small cellulose fibers which can thicken into strips or sheets. One example that offers great potential is the acetobacter xylinum species.

MATERIAL CONCEPT AND PROPERTIES

Bacterial cellulose is essentially finer than its plant-based counterpart and contains no disruptive additional substances. It is based on a highly complex, three-dimensional nanostructure, which gives the material outstanding mechanical properties. Particularly worthy of note is its high mechanical stability when wet, which is comparable to **Kevlar®** or **steel.** The bacteria strain used and the type of substrate together define the quality and thickness of the resulting sheet textile. It can be used in the human body and is colonized by the body's own cells, so allergic reactions are impossible. The properties of bacterially produced cellulose can be modified by changing the genetic code of the organism.

APPLICATION AND PROCESSING

Bacterial cellulose can grow into virtually any shape required and can be produced using a wide range of sugary substrates. During the manufacturing process it takes up a lot of water, which means it then needs to be dried, resulting in not inconsiderable shrinking. With its very good biocompatibility and high level of purity, the fiber is particularly interesting for medical and cosmetic purposes. The material can therefore be used as a wound covering, a hollow implant for bypass operations, or for soft part replacement. The lengthy growth period, as well as the price of around a hundred times that of plant-based cellulose, has hindered widespread industrial application to date. The particularly long fibers are also ideally suited to the production of high-quality paper. Applications as a substrate material for OLEDs or as a matrix for electronic paper are therefore being discussed. As far as foodstuffs are concerned, bacterial cellulose is known from the nata de coco dessert. A few years ago, media company Sony brought to the market membranes made of bacterial cellulose for headphones.

"Xylinum" stool with a surface made of bacterial cellulose (Design: Jannis Hülsen)

Item of clothing made from bacterially produced cellulose (Design: Suzanne Lee)

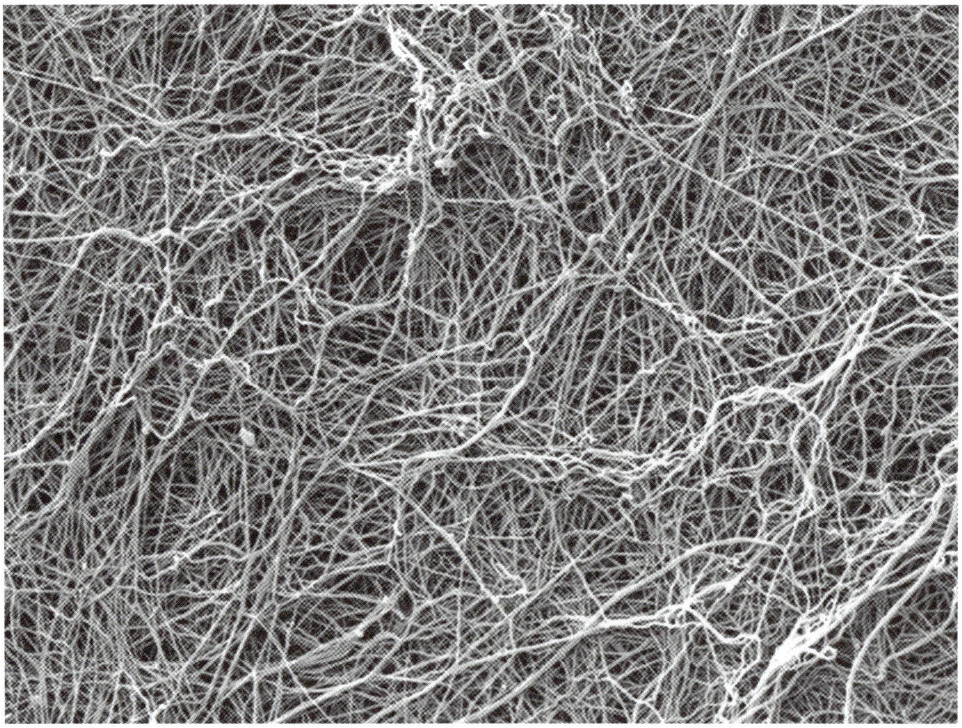

Bacterial cellulose on the nano scale (Source: fzmb)

Among designers, too, this extraordinary fiber material is extremely popular. At the Central Saint Martins College of Art and Design in London, for example, Suzanne Lee is looking into how microbes can enable entire items of clothing to grow organically. To achieve this, bacteria are cultivated in a bath of kombucha tea and sugar water, and in two to three weeks a gel-like sheet textile is produced. After drying, this feels like a plant membrane and can be composted entirely. In another project, designer Jannis Hülsen tested the potential of bacterial cellulose for furniture design. He designed a stool with a surface made from bacterial cellulose. After the final drying process, a feel and look comparable to parchment emerges. **Alder wood** has proved to be the ideal basis for this due to its very free structure.

One of the best known protein-based biopolymer is gelatin, which continues to be used in certain products such as Gummy Bear sweets or photographic paper.

Properties protein-based biopolymer // gel-like consistency // swells on contact with water // non-heat-resistant // no allergic reactions // high resistance to tearing

Sustainability aspects based on renewable raw materials // partly biodegradable

GELATIN

MATERIAL CONCEPT AND PROPERTIES

Gelatin is made from animal proteins (**collagen**), which are mostly found in the connective tissues as well as in the skin and bones of cows and pigs. Its special properties make this biopolymer particularly suited to the fields of foodstuffs and medicine. For example, it swells in water and dissolves at a temperature of 50 °C. What is particularly striking is its gel-like consistency, which is used for products such as **aspic,** jelly, and **liquorice.**

APPLICATION

Due to its low heat resistance, gelatin is not widely used in an industrial context. Further applications in the food industry are in ice cream, marshmallows, candy, and yoghurt. Its good biocompatibility makes gelatin suitable for the production of capsules and as a binding agent for tablets. Furthermore, it is used for coating implants and photographic paper, and in cosmetics the bioplastic can be found in creams and lotions. Gelatin substances are used predominantly as sheet materials for baking and cooking. The rise of digital cameras has meant gelatin has lost its importance as a material used for film in the photographic industry.

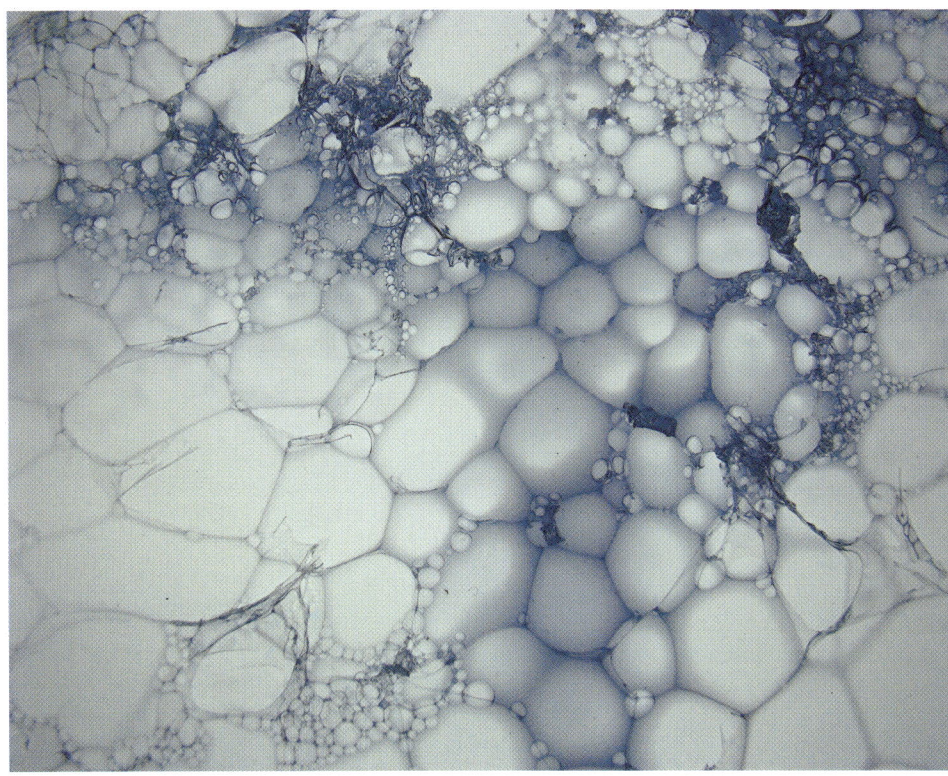

Bioplastics based on gelatin
(Source: CAAD, Manuel Kretzer)

Sources of gelatin
(Source: CAAD, Manuel Kretzer)

Gummy Bears made using gelatin

Gelatin sheeting for cooking and baking

Keratin forms the basis for the formation of fingernails, animal horns, spines, feathers, and hooves. It is a fibrous protein of animal origin, which can also be used as a natural bonding agent.

Properties protein-based biopolymer // insoluble in water // resistant to bacteria // resistant to strong temperature fluctuations // reduced formaldehyde burden in internal spaces

Sustainability aspects based on organic waste // biodegradable

KERATIN

MATERIAL CONCEPT AND PROPERTIES

No substance based on **keratin** dissolves in water. Keratin is also resistant to bacteria and to strong fluctuations in temperature. The mechanical properties depend on the relevant fiber structure of the keratin. The number of so-called **disulfide bridges** has a decisive influence on the resulting fiber resistance. The burning of keratin produces an unpleasant odor. Scientists have been able to prove that keratin fibers reduce the **formaldehyde burden** in an internal space.

APPLICATION

The use of animal hair in the field of textiles is a given. More unusual, though, is the use of hair, and particularly the strongly insulating hair of camels, as a filler material for sleeping bags. One example comes from southern Germany, where a producer of outdoor articles has launched a sleeping bag with a filler made of curled **camel hair,** moisture-absorbing **cellulose fibers,** and recycled polyester fibers. While the animal hair provides the filler with outstanding insulation properties, the polyester promotes the transportation of moisture and its exchange with the external environment.

Keratin adhesives are made using the fiber proteins from animal horn tissue and are used wherever other keratin-based materials need to be glued to one another. They are hard and resilient, long-lasting and entirely transparent. Keratin glues are popular for sticking real hair and are available on the market as granulate or in rods. The latter are generally used in heat guns.

Today camel hair is used as sleeping bag filling, among other things (Photo: Diana Drewes)

Berlin-based sculptor Iris Schieferstein uses predominantly animal remains in her art. She is known in particular for her shoe designs, in which she uses cow and horse hooves. Her clients include the internationally renowned singer Lady Gaga, who combined the shoes with a dress made of animal flesh as an outfit for an awards ceremony.

Horn comb composed of keratin

Shoe design using horse's hooves (Design: Iris Schieferstein)

Animal hair and its properties

Wool from a merino sheep	fine, soft, very curly
Camelid hair from alpaca, ilama, vicuña, guanaco	soft, shiny, fine, less curly
Hair from an angora rabbit	very fine, extremely light, smooth
Camel hair	very fine, soft, slightly curly, beige-brown
Mohair from angora or mohair goat	long, slightly curly, shiny, barely felts
Cashmere hair from a cashmere goat	fine, soft, lightweight, shiny
Hair from a yak	coarse
Horsehair	very coarse

Properties protein-based bioplastic // antibacterial // temperature-regulating and skin-smoothing // good moisture management

Sustainability aspects no chemical additives // use of non-tradable milk

MILK PROTEIN FIBERS

A new fiber for allergy sufferers has appeared on the market, which is claimed to have no negative effects on the skin. This is because, unlike conventional yarns, the manufacture of milk protein fibers does not require chemical additives.

MATERIAL CONCEPT AND PROPERTIES

Originally this casein protein fiber, which is also known as **milk silk,** came from Asia. Now a German company has entered this promising market. Milk protein fibers contain up to 18 **amino acids,** which support cell growth and prevent a reaction from the skin. They stimulate blood circulation, counteract itching, and smooth the skin. The moisture management of the **functional fibers** also prevents the growth of bacteria by 99% and promotes **temperature regulation,** which is important for allergy sufferers.

No milk is used in the fiber production that would also be suitable as a foodstuff. The feedstock therefore includes, for example, **colostrum milk** from cows just about to calf or the centrifuge waste from cheese production. Every year, around 1.9 million liters of non-tradable milk is disposed of by companies.

APPLICATION

With their complex profile of properties, milk protein fibers are useful not only for clothing, but also for the automobile industry and medical technology. Examples include antibacterial household textiles, bed linen with a cooling effect, heat-insulating car seat covers, and hygienic membranes for applications in medical technology. Even small quantities of milk protein fibers added to textiles can incorporate positive characteristics. Series production by the German manufacturer Qmilk began in mid-2013.

Colostrum milk as a basis for fiber production (Source: Qmilk)

Milk protein fibers (Source: Qmilk)

Milk protein fibers in production (Source: QMilk)

Milk protein fibers under the microscope (Source: QMilk)

GLYCOPROTEINS

Properties positive influence on skin // high viscosity

Sustainability aspects based on renewable raw materials // partly biodegradable

Some natural substances and particular green algae have a favorable effect on human skin. Scientists in Taiwan have identified that even eggshell membranes can be beneficial to the healing process of wounds.

MATERIAL CONCEPT AND PROPERTIES

The eggshell membrane mainly consists of **glycoprotein,** i.e. macromolecules made up of a protein and several sugar groups. Glycoproteins fulfill various roles within the organism, for example they function as a lubricant in mucus and can be found in cell walls as a structural element. Solutions with glycoproteins boast high viscosity.

APPLICATION

The structure of glycoproteins shows similar potential as regards opportunities for using milk proteins as a fiber material, or bacterially produced cellulose as a leather-like textile in fashion. Designer Ulrike Böttcher has looked at various ways of using eggshell membranes for this same purpose. The possibility is still very much in the development stage, however.

Eggshell membranes
(Source: Ulrike Böttcher)

SPIDER SILK PROTEINS

Properties protein-based bioplastic // high elasticity // extremely tear-resistant // good moisture management

Sustainability aspects sustainable producibility // suitable for recycling

Spiders have inhabited the earth for around 400 million years and have developed a wide variety of ways to catch their prey, one of the best known being the spider's web. The fibers and webs that spiders produce in the open air boast a unique level of stability and elasticity. Scientists have now succeeded in recreating the proteins from spider silk in a fermentation process with genetically modified bacteria and are manufacturing them on an industrial scale.

MATERIAL CONCEPT AND PROPERTIES

Spider silk boasts outstanding tearing strength which, in relation to its delicate structure, exceeds that of steel. The material absorbs three times as much energy as **nylon** or **Kevlar**® and, moreover, it is as elastic as rubber. This combination of material properties is not achieved by any other fiber material. In addition to its mechanical properties, spider silk boasts characteristics that are useful for medical applications, as it does not trigger any allergic reactions in the body and can be used without fear of harm. **Spider silk proteins** can be applied to virtually all synthetic and natural materials and improve mechanical properties as a coating.

APPLICATION

Industrially-produced yarn made from recombinant spider silk protein was launched in the spring of 2013. It shows similar characteristics to natural spider silk fibers. In addition to the thread format, technologies are already available to process proteins as a raw material into balls, membranes, films, and foils. AMSilk offers cell culture sheets that are given a thin **silk coating** or that contain an open-pore foam matrix made of spider silk. Now, inserts made of spider silk fleece material are also available. Due to the excellent tolerability of spider silk proteins by the human body, there is potential for their use in cosmetics, medicinal implants, and as a suture material. Currently, research is being carried out at Hanover University Medical School into the use of silk from the golden **orb-weaver spider** of Tanzania to form artificial skin.

Spider's web
(Source: University of Bayreuth)

Biosteel® fibers produced from recombinant protein boast similar properties to natural spider silk fibers (Source: AMSilk)

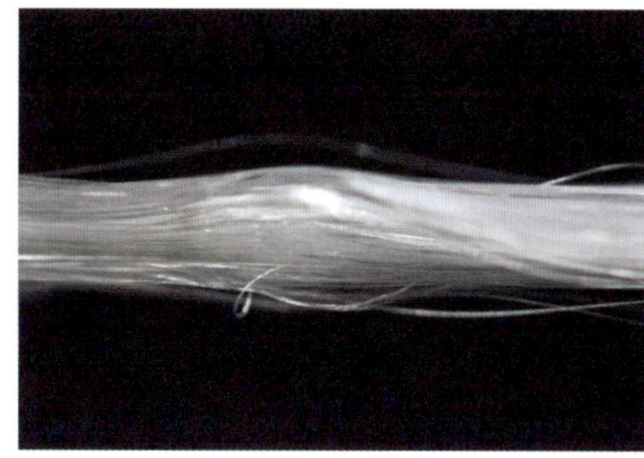

Artificially produced spider silk
(Source: University of Bayreuth)

SOYA PROTEIN FIBERS

Properties protein fibers from plant origins // glossy, smooth, and soft // quick to dry // temperature-regulating effect // antibacterial // stores heat

Sustainability aspects biodegradable // waste product from the food industry // based on renewable raw materials

Although soya protein fibers were developed as far back as the 1940s, their importance has only grown with the increasing scarcity of cotton fibers and the orientation towards sustainable clothing in recent times. The commonly used term is soya silk, which refers to what is currently the only known protein fiber of plant origin. It represents a basis for vegan fashion.

MATERIAL CONCEPT AND PROPERTIES

This fiber material's glossy, smooth appearance and its soft feel immediately suggest a similarity to silk. The natural color varies between ivory, caramel beige, and golden yellow. Soya proteins arise from the extraction of **soya oil** or from tofu production. They are a waste product of food production and are entirely biodegradable. No additional cultivated land is required for their production so there is no negative impact on food prices. Soya protein fibers dry out quickly, help to regulate temperature and have an antibacterial and antifungal effect, preventing fungi and bacteria from accumulating. Their good heat-storing properties are comparable to those of cotton.

APPLICATION AND PROCESSING

Soya silk is conventionally used in textile products. Soft yarn made of soya protein fibers is used in knitted garments in particular. Most importantly, it is suitable for people with **wool allergies.** Soya silk is combined with other fibers when processed. The Umasan label, founded in Berlin in 2009, specializes in the production of vegan fashion. It uses no animal-based fibers and relies on materials such as cellulose fibers with enriched algae minerals, soya silk, or knitting yarn made of seaweed.

Vegan fashion with soya silk
(Source: Umasan)

Properties quickly growing biomass // three-dimensional growth in water // varied potential for use in material hybrids

Sustainability aspects based on renewable raw materials // no agricultural land needed

ALGAE-BASED PLASTICS

Algae with names like kombu, hijiki, nori, and wakame are familiar to us from the food industry. They come from various different waters and in diverse colors and sizes. Unlike land plants, they do not require cultivated land to grow, so it is hoped that these quickly growing biomasses will offer new sources for energy provision or the production of plastics. Algae can also form the basis for the production of gelling agents and impression materials. Some designers have already begun working on designing products using algae.

Algae on the beach in Namibia
(Source: VPZ Graz)

MATERIALS

Algae fiber-reinforced thermoplastics

Algae are already used industrially as a strengthening material for plastics. In 2012, the very first thermoformable hybrid material made of 80% polypropylene and 20% algae was presented in the USA. In addition, the company Algix has developed formulations with PE or ethylene vinyl acetate (EVA) and the bioplastics PLA, PHA, or thermoplastic starch (TPS). These can be injection molded to form structural elements, extruded to form films, or spun to form threads.

Algae fiber-reinforced plastic granules (Source: Algix)

Film made from algae fiber-reinforced thermoplastic with addition of algae (Source: Algix)

Fibers and films made of algae oil

In the European SPLASH (Sustainable Polymers from Algae Sugars and Hydrocarbons) research project, which runs until 2016, scientists and companies are working to develop procedures for the manufacture of algae-based monomers as a basis for the plastics industry. The aim is to establish processing chains for the production of polyester for packaging and consumer items, or fibers for yarns and meshes. In relation to this, researchers at the Fraunhofer UMSICHT are working on the biotechnological extraction of algae oil from the green algae **botryococcus braunii** by microorganisms.

CARBON DIOXIDE POLYMERS

Properties carbon dioxide as a raw material // substitution of conventional plastics possible // processing using standard techniques

Sustainability aspects capture of the greenhouse gas carbon dioxide // bioplastic in combination with organic waste

Carbon dioxide is one of the key elements of the atmosphere, and the increasing volume of it has been blamed for the climate warming of the last few years. The consensus among scientists is that around 60% of the anthropogenic greenhouse effect is related to CO_2 emissions. The reduction of emissions and the capture of free carbon dioxide are therefore the subject of a comprehensive series of research projects. Some scientists are even trying to use it for the production of plastics.

MATERIALS

ABS substitute based on carbon dioxide

Siemens and BASF are currently developing a plastic compound as an alternative to the bulk plastic acrylonitrile butadiene styrene (ABS) that is made up overwhelmingly of renewable raw materials and carbon dioxide. The new compound has similar properties to ABS, which is a variant of polystyrene, yet has a significantly better ecological balance. It is based on a mixture of rough **polyhydroxybutyrate** (PHB) and **polypropylene carbonate** (PPC) as plasticizing agents. The PHB used in the process is made from renewable raw materials such as palm oil or starch. The PPC consists of up to 43 weight-percent carbon dioxide, which comes from power station waste gases. The new material is biodegradable, light-permeable, and can be processed using conventional technologies. At Bosch-Siemens Household Appliances, it has already been used to make a vacuum cover under series conditions.

Carbon-dioxide-based PUR

A development team at Bayer MaterialScience is working on the implementation of a pilot system to convert carbon dioxide into PUR. This is processed in large quantities to create foam materials for mattresses, fridges, or building insulation. In a new catalyst process, CO_2 from the power industry is chemically bonded and replaces a portion of the petroleum previously required for PUR production. Collaborating with Bayer are the energy group RWE, the RWTH Aachen University, and the CAT Catalytic Centre. Industrial production is planned for 2015.

Vacuum cleaner cover made of carbon dioxide polymers (Source: BSH)

Pilot system for the production of PUR with carbon dioxide (Source: Bayer MaterialScience)

The limonene found in orange peel can be used as a basis for bioplastics.

Bioplastics with carbon dioxide and orange peel

Another focus of current research is the development of a bioplastic based on carbon dioxide and **orange peel.** It was some years ago that American scientists at Cornell University first succeeded in making the **limonene** contained in orange peel react with carbon dioxide to produce a plastic called **polylimonene carbonate,** which boasts properties similar to polystyrene (PS). The limonene molecules, which consist mainly of carbon, can be found in around 300 plant species and are used most frequently as a fragrance in cleaning products and detergents. The reaction of limonene with oxygen produces limonene oxide, which does not normally react with carbon dioxide. The scientists used a specific catalyst to trigger the reaction. In this way, they were able to create the long molecule chains made of carbon dioxide and limonene oxide units, which are important for the plastics industry. Scientists from Freiburg's Material Research Center have also now mastered the process without having to use solvents. In the procedure developed in Freiburg, carbon dioxide is chemically bonded to limonene oxide. The resulting **limonene dicarbonate** can be molded and converted to PUR with so-called **amine hardeners.**

Properties chain-like molecular structure // water-soluble // thickening and water-binding properties // highly swellable // lubricant properties

Sustainability aspects based on renewable raw materials // biodegradable

STARCH ADHESIVES

Since many product designers eagerly await the development of biodegradable products, natural bonding agents based on starch are becoming increasingly important. In the EU, starch is predominantly extracted from corn, potatoes, or wheat.

Potato starch as a white powder (Photo: Diana Drewes)

MATERIAL CONCEPT AND PROPERTIES

Starch consists of **polysaccharides** and has a chain-like molecular structure in parts. It is present in all **chlorophyllaceous plants** (for example, rice, wheat, and corn), is formed during **photosynthesis,** and serves to store energy. Its use in the human body requires **enzymes,** which break down the starch molecules into smaller components. Starch is available as white powder and can be dissolved in water. It has adhesive properties as well as thickening and water-binding qualities and can thus influence the flow characteristics of fluids (**rheology**).

APPLICATION

In addition to uses in the food industry for the production of sweets or baked goods, for example, starch is used as a bonding agent for tile adhesives, concrete mixes, corrugated cardboard, gypsum concrete, and chipboard. In the past, flour or starch was boiled in water to produce wheat paste. Other applications include lubricants, plastics, films, and tablets, for which it can also be used as a thickening agent, a carrier, and a filler. A not insignificant proportion of national starch production is used by the paper and chemical industry. **Artificial snow** can also be produced from starch, and biofuels such as **bioethanol** are mainly based on it. While in Germany starch tends predominantly to be produced from wheat, in Brazil **sugarcane** is the most common source.

Potato starch

Beside its 21% water content, a potato consists entirely of starch. In order to extract the starch, the potato is first chopped, triturated, and washed. As a consequence, the **starch grains** rupture and, after the subsequent drying process at temperatures under 60 °C, and a sieving procedure, what remains is a white powder that can swell substantially. This can be dissolved in warm water, initially thickening and expanding. With a temperature of 62.5 °C a starch adhesive is produced which is suitable for various applications. **Potato starch** is also used as a lubricant for the board game Carrom.

Corn starch

The stiffening capacity of **corn starch** is greater than that of potato or wheat starch. Some 71% of a corn plant is made up of corn starch. To extract it, the corn is softened in warm water at 35 °C, then washed, dried, and ground. The resulting meal is sieved until a fine powder remains. Corn starch can be used both as a bonding agent for handicrafts and as an environmentally friendly paste glue. **Glycerin** can be added to improve its spreadability.

Wheat starch

In order to extract **wheat gluten,** thin-husk, mealy grains are softened in water, ground in a mill, and washed in various different phases. The fibers and husks are removed during the process. Upon contact with water and at a temperature of 67.5 °C, wheat starch takes on adhesive properties. Wheat gluten is used as a bonding agent in **wood pellets,** for example. It is particularly good for bonding paper and wood surfaces.

Rice starch

At 89%, rice has a particularly high starch content. **Rice starch** is used in textile production to help fabrics maintain their shape and to give them dirt-repelling properties. Boiled up in quantities of 40 grams to one liter of water, rice starch can be used as a biological bonding agent for gluing paper, cardboard, and textiles.

Starch content of certain plants

Plant	Starch content
Rice	89%
Potatoes	82%
Barley	75%
Sorghum	74%
Wheat	74%
Sweet potatoes	72%
Rye	72%
Corn	71%
Peas	40%

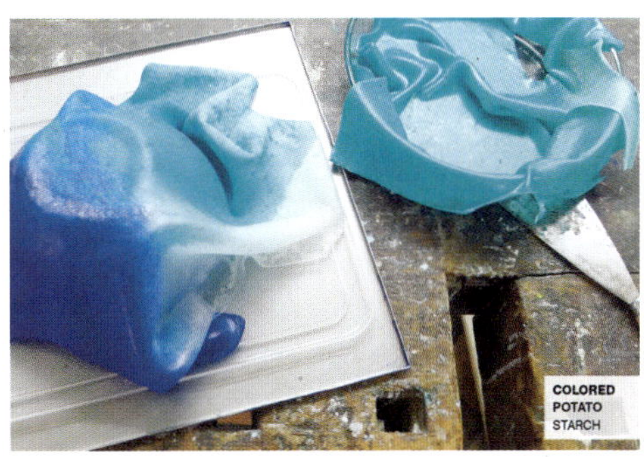

Bio-art made of dyed potato starch
(Source: Eva Marguerre, Marcel Besau)

Potato starch paste with almond scent
(Source: Modulor)

Japanese starch paste based on tapioca
(starch from the manioc root)

COLLAGEN ADHESIVES

Properties animal origin // thermoplastic properties // modification of properties possible through use of additives // high elasticity // lubricant properties

Sustainability aspects based on renewable raw materials // biodegradable

In addition to plant-based natural adhesives there are also biological bonding agents that come from animal sources. These include bone, fish, and hide glue. Their main component is collagen, the protein found in gelatin, and they are known as animal glues.

MATERIAL CONCEPT AND PROPERTIES

Animal glue is obtained by boiling up animal products such as **fish bones** or scales, swim bladders, cartilage, **bone, animal skins,** or rabbit fur. The result is a viscous mass which hardens when cooled. Animal glues react thermoplastically, becoming soft under heat and thus malleable. The strength of the bond is improved by pressure and the processing time is generally relatively short. Additives can modify their processing properties and adhesion strength. **Tannin** and **alum** make animal glues insoluble, while **alcohol** makes them better able to penetrate pores. **Acetic acid,** by contrast, reduces the melting temperature and increases bonding strength by up to 20%, and the addition of glycerin can give the glue elastic properties. **Zinc sulphate** is used as a preservative.

APPLICATION

Animal glues are being revisited by designers in the development of biodegradable components. They are used industrially as a bonding agent in the wood and paper industries. In the construction industry, too, they have a few specific uses. Animal glue is suitable, for example, as a delaying agent in plaster of Paris and is useful in the preservation of monuments in particular. Furthermore, it is used in inks and photographic paper. Another typical area of use for animal glues is in musical instrument manufacturing.

Bone glue

Bone glue, which is used most frequently for the restoration of old furniture, is obtained either through boiling or steaming animal bones and cartilage. When used on wood, a crystal-clear adhesion layer is created. The product used to be known as **joiner's glue** and was supplied in three grades of brightness. **Bone glues** have only a faint smell and are nowadays used in the field of art, for the production of antique-style picture frames, or for **inlaying.**

Hide glue

Hide glues are now used wherever a slightly elastic joint is required as part of a nonindustrial process. Examples include **bookbinding** and **canvas priming.** The elasticity of the glue allows wooden instruments to vibrate more freely and is conducive to the **sound formation**. In the past, hide glue was used on the reverse of **postage stamps.** Nowadays it is mainly produced from **cattle hides** and sold in the form of granules or pearls.

Fish glue

This natural glue is obtained by boiling up **fish skins** and solid **fish waste products**. At room temperature it remains in liquid form. What is striking in comparison with other animal glues is its good adhesion on metal and ceramic surfaces, but users often complain of an unpleasant odor. **Fish glues** are used for musical instruments, for gluing wooden parts together, and for inlays in furniture surfaces. One product particularly well known in the design world is the fish scale plastic by designer Erik de Laurens, which is based on the bonding strength of fish proteins. An exceptionally high-quality fish glue is obtained by using the **swim bladder** of the **sturgeon** as a base material.

Furniture inlays with fish scales
(Source: Erik de Laurens)

CASEIN ADHESIVES

Casein is a mixed substance that comes from the proteins in milk. It provides young animals with important nutrients for growth, such as phosphorus and calcium. In cheese and quark, casein forms the main constituent of the proteins and is a basis for curdling to form a solid substance. This is why casein can also be used as a basis for adhesives. Indeed until the 1930s the plastic galalith was obtained from casein.

Properties based on milk proteins // excellent adhesion // moisture-proof // processed cold

Sustainability aspects based on renewable raw materials // biodegradable

MATERIAL CONCEPT AND PROPERTIES

The importance of **casein glues** for furniture-making goes back to ancient times. To produce the glue, the casein must be swollen in water and macerated with alkaline compounds such as **lime** or **borax.** The reason for this is that the casein contains **phosphorus,** which makes it insoluble. For nonindustrial use, casein glue can be produced with **quark** and lime in a volume ratio of five to one. After being broken down, casein glues boast excellent adhesion, can be processed cold, and are moisture-proof.

APPLICATION

While in the past casein glues were frequently used as bonding agents in **mortar** or **wood-based materials,** they have lost their former significance in many areas with the advent of synthetic adhesives. Nowadays they are still used as bonding agents in inks or as **label adhesives** or are used for gluing **linoleum,** a highly resistant cover material for floors and work surfaces. This is entirely due to their natural base materials such as **linseed oil, natural resin,** and wood and **lime powder** which, in combination with the bio-glue, make them potentially biodegradable.

SOYA ADHESIVES

The proteins contained in the soya plant can be used to produce a biological glue with strong adhesive properties.

Properties protein-based // normally water-soluble // desensitization-to-water modification through unfolding of protein molecules

Sustainability aspects based on renewable raw materials // free of formaldehyde // biodegradable // small carbon footprint

MATERIAL CONCEPT AND PROPERTIES

Soya adhesives are free of formaldehyde and are biodegradable, representing a natural alternative to synthetic adhesives. They also have a smaller **carbon footprint** compared with petrochemical adhesives, smaller even than that of **casein glue.** Like other bio-based adhesives, however, one problem when it comes to potential applications is water solubility. So scientists have developed a water-resistant modification. This involves a process whereby the **protein molecules** are unfolded in a certain way, so that the hydrophobic amino acids are directed outwards.

APPLICATION

At present, soya adhesives are still only rarely used. Most significantly, they offer the potential for indoor applications, for example in furniture or as parquet adhesive. Some manufacturers are therefore already offering biocomposites and fiber-reinforced sheets that use soya proteins as a bonding agent. The desensitization-to-water modification, in particular, makes them appropriate for wood-based materials and plywood. Plywood made using soya adhesive and cherry, pine, or walnut woods was tested with three soaks in a water bath (each for 48 hours) and subsequent drying, and revealed virtually no dissolution and a reduction in overall adhesion of just 10%.

56

BIOPLASTICS AND
BIO-BASED BONDING AGENTS

Biocomposite made using a soya adhesive
(Source: e2e Materials)

Mussels have an adhesive that enables them to grip to virtually any surface with strong adhesion even underwater. As a result of these properties, the biocompatibility, and certain additional functionalities, researchers all over the world are working on reproducing mussel adhesive and exploiting its potential in an industrial context.

Properties adhesion properties even underwater // hardens when pH value changes // loss of adhesion under UV light // biocompatible

Sustainability aspects strong and long-lasting adhesion // self-healing properties // recyclable

MUSSEL ADHESIVES

MATERIAL CONCEPT AND PROPERTIES

The strong adhesion of **mussel glue** on virtually all surfaces such as rock, wood, and metal is achieved with specific proteins that the **shellfish** get from **amino acid dihydroxyphenylalanine** (DOPA). In the ocean, extremely stable, densely cross-linked polymer compounds are formed, which have a strong adhesion effect on inorganic oxides in rock. Mussel glues are also able to absorb **metal ions** from seawater and reproduce the polymer structure by themselves. These self-healing properties are particularly interesting when it comes to serious safety applications. Currently in the development stage are mussel glues that harden when there is a change in **pH value,** and lose their adhesive properties and dissolve again under the influence of UV radiation.

APPLICATION

Mussel glues offer great potential for innumerable technical and medicinal uses. They are equally appealing for use in underwater pipelines as they are for **self-healing wings** on aircraft. Their high biocompatibility also means that mussel glues lend themselves to use in the human body, for securing implants, sealing bleeding wounds during an operation, or closing up the amniotic sac in pregnant women.

Mussels

Lignin is a biopolymer contained in the cell walls of all plant fibers, which serves as a hardening component in the cell tissue. Its thermoplastic properties make lignin suitable for use as a bonding agent in various different industries.

Properties thermoplastic properties // physiologically harmless // insoluble in water // brittle // activation with enzymes possible

Sustainability aspects based on renewable raw materials and waste products // biodegradable

LIGNIN

MATERIAL CONCEPT AND PROPERTIES

Lignin boasts good mechanical stability, is physiologically harmless, but is also insoluble in water and therefore more difficult to break down biologically than other natural substances. The biopolymer is brittle, absorbs ultraviolet light completely and is brownish in color. Lignin can be obtained biologically using bacteria and fungi or can be macerated in technical chemical processes. In the **paper industry,** types of lignin can be found as waste products in spent liquor, with varying molecule sizes, in the form of **kraft lignin, lignosulfonate,** or **organosolv lignin.** The variety of the lignin types and the impurities in the waste water mean that this waste product is generally used for energy purposes only.

Green lamp made of Arboform® with a lignin matrix. Due to the low shrinkage, it is possible to produce significant fluctuations in wall thickness. (Design: Romolo Stanco, source: Tecnaro)

APPLICATION

In some areas, however, lignin has proved itself as a bonding agent and additive. It can be used as a putty, for example, which holds the shape of particles in wood pellets after the heat and pressing process. It is also now used in the manufacture of wood-based materials as an alternative to potentially harmful synthetic resin. It is the main component for Arboform® as well, a bioplastic which is suitable for injection molding and extrusion. **Lignosulfonates** can be used too as **form sand binders** in casting and for improving flow in the construction materials industry. In the construction of nonasphalted roads, lignin helps to prevent the formation of dust. Although the potential for application is vast, the various types of lignin are still used relatively rarely. However, scientists expect to see a big rise in the significance of lignin bonding agents, particularly for the wood industry, and are developing procedures to replace synthetic adhesives.

Binderless wood-based materials

At the Georg August University in Göttingen, scientists have developed a process for the manufacture of binderless wood-based materials. To achieve this, they made use of the self-adhesive strength of the lignin found in wood. The lignin molecules are activated by **fungal enzymes,** which are already converted on a large scale for other applications. Applied to grind up cellulose fibers, these stick to one another independently, without the need for an additional bonding agent.

BIO-BASED RESINS

Resins are viscous substances that harden under certain circumstances to create thermosetting masses. In addition to the synthetic resins based on polyester, epoxy, polyurethane, phenol, or silicone, which are already well known within the industry, there are also natural examples.

Properties viscous // not uniform substances // harden upon contact with the air // brittle

Sustainability aspects largely coming from trees // can decompose in a natural way

MATERIAL CONCEPT AND PROPERTIES

Plant resins generally come from trees. They are secreted to seal up tears in the bark and to protect the tree from outside influences. Natural resins are not chemically a uniform substance, but rather a mixture of **resin acids** and **terpenes.** They harden into a brittle material once they leak out and come into contact with elements in the air that become volatile after evaporation, and are insoluble in water. Conifers tend to produce more resin than deciduous trees. The plant resin most frequently used in adhesives or chewing gum, for example, is **colophonium.** It is obtained from the resin of pines or spruces and is left over as a solid component after distillation of **turpentine oil. Amber** is also the fossilized resin of a spruce from prehistoric times. The only known example of animal resin is **shellac.**

APPLICATION

Compared with their synthetic counterparts, natural resins play only a small role. Industrially they are used as an additive in the production of adhesives, varnishes, and inks. In the context of the increasing orientation of our product culture towards natural materials, however, bio-resins are also gaining in importance among designers.

SHELLAC

Properties animal origin // harmless to health // colorless // hard and brittle // swells upon contact with water // stable up to 100 °C

Sustainability aspects based on excretions // biodegradable

Shellac has long been familiar to us as a surface coating for old records. In a large number of applications this material, produced by the lac insect, has been replaced by plastics generated using petrochemical processes.

MATERIAL CONCEPT AND PROPERTIES

Shellac is a biodegradable and harmless substance obtained from the excretions (resin) of the **sternorrhyncha.** At room temperature, the natural resin has a hard, brittle consistency with a silky shine, which makes it particularly suitable as a coating material. Upon contact with water the material swells, but cannot be dissolved. It behaves differently, however, with alcohol, **ammoniac,** or **borax.** At temperatures of 100 °C or more, shellac begins to decay and, in doing so, gives off an unusual odor. In general, shellac is colorless, but can also appear yellowish.

APPLICATION

Shellac can be used as a biological alternative to conventional varnish or as a **polish** for wooden items. In art and graphics, shellac is known predominantly as a bonding agent in paints. It is also used in varnish production for musical instrument manufacture and as a releasing agent in plaster molds. In the food industry, it is used as a coating for chocolate. Shellac is also found in hairspray and as sealing putty for the electronics industry. It is the typical substance used for coating blackboards, and is also useful for fixing design components made of paper pulp.

Shellac surface for the fixing of molded parts made of paper pulp (Source: Svenja Bechtel)

NATURAL WAXES

Wax is another substance often used in connection with the manufacture of moldings in the design world. It is predominantly of interest given its low melting temperature.

Properties low melting temperature // no exact melting point // water-resistant // sticky surface // fatty acids influence their consistency

Sustainability aspects natural origin // biodegradable

MATERIAL CONCEPT AND PROPERTIES

The term **waxes** refers to a group of materials of differing chemical compositions, which are solid at room temperature, water-resistant, and have a sticky surface. While most waxes can be kneaded at room temperature, **hard wax** must be warmed slightly to become malleable. Particularly striking is the shine created when the surface is rubbed. Depending on the quality of the wax, it will start to melt at a temperature of 40 °C. There is no exact melting point here, but scientists talk about a temperature bracket in which wax becomes liquid. The quality and the type of the **fatty acid** contained in the structure have a decisive influence on the consistency and the properties of the wax when subject to temperature changes. Two-thirds of all natural waxes originate in plants. Synthetic wax with precisely defined properties is, by contrast, produced using crude oil.

APPLICATION

Up until the 18th century, candles were produced almost exclusively from beeswax. Today, waxes are a component in furniture polishes, wax coatings for wood, creams and lotions, as well as in automobile care or impregnation agents. They are increasingly being used in the context of the development of biodegradable design and product prototypes. One example is the design for compostable flower pots made of a purely plant-based fiber composite, coffee grounds, and natural waxes as a bonding agent, which was awarded first prize in the adream 2012 sustainability competition.

Pine wood impregnated with wax (Source: Dauerholz)

Flower pots made of coffee grounds (Design: Sanam Viseux)

Natural waxes

Animal waxes	beeswax, chinese insect wax, shellac wax, spermaceti, wool grease
Plant waxes	carnauba wax, candelilla wax, flax seed wax, sugarcane wax, coffee wax, rubber wax
Fossil waxes	paraffin, montan wax, peat wax, sapropel wax

Synthetic waxes

myricyl palmitate, butyl stearate, cetyl stearate, hydrocarbon waxes, cholesteryl palmitate, diglycol stearate

Properties yeast cultures in a binding function // sensitive to moisture // properties like natural stone following sintering process // food-safe

Sustainability aspects Moldings produced using quarry waste // binder based on renewable raw materials

YEAST CULTURES FOR MALLEABLE STONE

Given its high level of stability and hardness, granite is extremely popular as a material for flooring or wall paneling. Now a manufacturing process has been developed with the use of yeast culture, whereby granite can be molded three-dimensionally like ceramics.

MATERIAL CONCEPT AND PROPERTIES

Granite waste from quarrying is initially broken down, ground up in mineral grinders, and then mixed with active water. This contains **yeast cultures** which, in a so-called **"aging" process** lasting several days, are bound and grown together with the **granite grains.** This results in a malleable mass that can be processed in a similar way to ceramic clay. Technical expertise is needed, however, since the **granicium** combined with effective **micro-organisms** reacts very sensitively when moist. The molded components are reinforced under high pressure in hydraulic presses and are dried for two weeks. The final production process and fusion of the **quartz particles** then follows in a kiln at a temperature of 1,300 °C. During this stage the yeast cultures burn up, resulting in a molding that has similar material properties to **natural stone.** It is extremely resilient, highly stable and food-safe, and boasts a gray color with the usual shading.

APPLICATION

This process can be applied to create moldings for accessories and kitchen utensils. The manufacturer has also already considered elements for outdoor furniture.

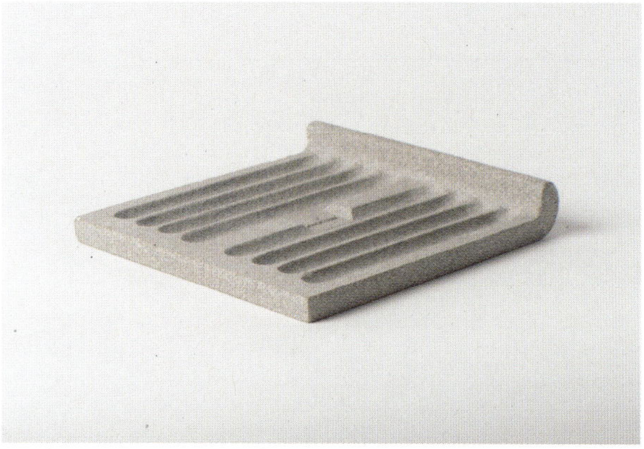

Molded part made of granicium using yeast cultures in the production process (Source: DENK Keramische Werkstätten)

Seating elements made of moldable stone (Source: DENK Keramische Werkstätten)

BIOBITUMEN

Properties quick to bond // good roadworthiness // biodegradable in case of erosion

Sustainability aspects partial redundancy of petrochemical products // no plasticizers // use of renewable raw materials based on organic waste and residues

Bitumen is a waste product from crude oil processing and is currently the most important bonding agent in road construction. The material has been used since ancient times, when it was deployed to protect the outside of ships from water penetration. The ever declining availability of crude oil has meant that the price of bitumen has multiplied over the last few years, prompting manufacturers to seek alternatives.

MATERIALS

Rape asphalt

Purely concrete roads cannot replace **asphalt** in the long term, since more energy is required for the production of cement. This has led to a development in Austria, where a type of asphalt has been launched that contains up to 20% **rapeseed oil,** so reducing the quantity of bitumen required. Furthermore, the asphalt mixing does not require soft bitumen solutions, and hard bitumen is prepared for its particular use through the addition of rapeseed oil. The quick binding nature of **rape asphalt** guarantees good roadworthiness leading to a reduction in road construction. The bonding with loose chippings is optimized. In addition, rapeseed oil is biodegradable in rain where erosion occurs. The product's good environmental properties are even greater since no plasticizers are required.

Molasses asphalt

In Australia, a **biobitumen** based on **sugarcane molasses** has been developed and proved to be functional. The bonding agent consists of low-molecular and water-soluble organic waste materials such as sugar, natural rubber (latex), tree resins, rubber **colophony resins,** lignin, cellulose, and vegetable **cooking oil.** The asphalt developed on the biobitumen produced by Ecopave Australia can be produced in thicknesses of 1 – 200 mm.

NATURAL MATERIALS AND ORGANIC WASTE MATERIALS

02

Spectacle frames made of fish scales, materials with eggshells, lampshades created from coffee grounds, and materials based on stinging nettles. Inspired by a resurging environmental consciousness in society, both designers and materials manufacturers are looking for alternative materials designed from natural substances. With extraordinary speed, companies are now concerning themselves with something that environmental organizations have been advocating for years, namely renouncing petrochemical products and thinking in closed cycles. In this context, it is above all the use of organic waste materials that is increasing in significance. Examples include decking boards made of rice husks, veneers formed from banana tree fibers, raw material consisting of residual products from sugar production, and chipboard made of wheat straw.

Developments in recent years, with designers conceiving edible packaging, furniture made of sugar, and luminaires derived from algae, demonstrate that many designers are increasingly seeking to create biodegradable and ecologically sound products, packaging, and furniture. There is even potential to produce electronic components from natural substances, and conductive circuits made of organic materials. Environmentally damaging processing technologies are increasingly being replaced by those based on natural substances. Rapid-growing plants like the water hyacinth, the cultivation of which has led to problems in some regions, are now being used as a source of fibers in product development and design.

NATURAL MATERIALS AND ORGANIC WASTE MATERIALS

NATURAL FIBER COMPOSITES AND UNUSUAL ORGANIC FIBERS

Properties lightweight solutions using natural fiber // replacement of metallic components by NFRP moldings // unlikely to split // good acoustic properties // typical wall thicknesses of 2.5–10 mm

Sustainability aspects based on renewable raw materials // positive ecological balance possible for semifinished production

Driven by the need for lightweight construction materials and components, the automobile industry in particular is progressively replacing metallic components with fiber-reinforced plastics. Fiberglass and carbon fiber in particular have proven themselves as reinforcement materials, but as organic materials and organic waste gain in importance as alternative materials for industry, developers increasingly have natural fibers at their disposal. However, while they have a wide range of ecological, mechanical, and economic advantages to impress product developers, industrial designers, and design engineers, natural-fiber-reinforced plastics (NFRPs) have so far failed to be adopted for widespread use.

MATERIAL CONCEPT AND PROPERTIES

For the manufacture of NFRP components, fibrous plant constituents (in Germany predominantly **hemp** and **flax fiber**) are combined with plastic as a matrix material and subsequently pressed in a mold. The weight proportion of the polymer lies between 30 and 50%, so the parts are significantly lighter than those produced conventionally. NFRP components based on bast and leaf fiber are less likely to split, which is an advantage for automobile construction in particular. Another striking feature of NFRP components is their good acoustic properties.

APPLICATION AND PROCESSING

In the last decade, technology has been exploited for interior paneling for doors and fittings, for example. As with other composite materials, the manufacturing process offers a wide range of design possibilities. The selection of the fiber material, the matrix plastic, the mix ratio, and the bonding process can all be adjusted to determine the material's qualities, spanning a broad range. Fiber moldings can be produced in various different sizes and are easy to punch and cut. Compounds with other materials can also be realized in a straightforward manner in a pressing process through insertion, stiffening, and lamination. The achievable component thickness of sheet parts is between 2.5 and 10 mm.

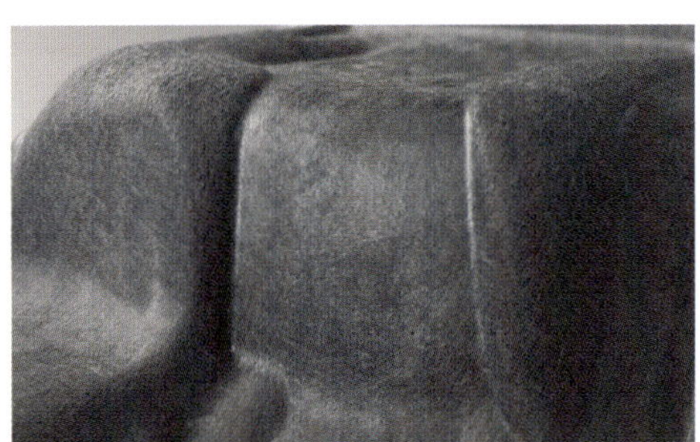

NFRP molding in the automobile industry (Source: Grimm Schirp)

Barktex® Fluffy – composite made of barkcloth with wool and a gauze made of cotton (Source: Barkcloth)

"Hemp Chair" based on a 75% proportion of hemp and kenaf fibers (Design: Werner Aisslinger)

UNUSUAL FIBER MATERIALS

Banana plant fibers

In Brazil, huge quantities of organic waste are produced every year, which have been used very little until now. As a result, a series of research projects is looking into how plant waste from Brazilian agriculture can be made available for industrial production. **BananaPlac** is the result of a project undertaken by the University of Rio de Janeiro which succeeded in creating a thermally moldable fiber material using **banana plants.** The fibers are made into a solid compound with biologically produced **polyurethane (PUR)** as a matrix material and pigmented in various different colors.

Bamboo hard fiber

Materials made of **bamboo hard fiber** are a possible replacement material for solid wood. The **bamboo strips** are crushed in the manufacturing process and the fibers are pressed with **phenolic resin** (weight proportion: 20 – 30%) under very high pressure. The resulting construction material is sufficiently robust to be used as a cover material for floors, walls, and ceilings. Bamboo hard fiber materials have also been used successfully in the Netherlands as static construction materials.

Coconut fibers

Coconut fibers are the world's only rot-resistant plant fibers. They are extracted from the shell of the **coconut** and are highly resistant to tearing and scrubbing. Particularly notable are the very good heat-insulating properties, given the many air pockets they contain. Since the fibers have a positive effect on the microclimate, they are frequently used in mattresses. Coconut fiber panels can also be used for **noise insulation** and for **protection from microwaves.**

BananaPlac board
(Source: Barkcloth)

Bulrush fiber components with protein glue
(Source: Naporo)

Bast fibres	Seed fibers
Hemp fiber	Poplar fluff
Bamboo fiber	Cotton made from the seed hairs of the fruit of the cotton plant
Stinning nettles	Kapok made from the inside of the capsule fruit of the real kapok tree
Jute	AKON
Kenaf	
Linen	**Fruit fibers**
Hops	Coconut fibers from the pericarp of the coconut
Ramie	Banana plant fibers
Abacá hard fibers	Hazelnut and peanut fibers
Pineapple	
Neptune grass	
Sisal from agave leaves	

BananaPlac
(Source: Barkcloth)

Technoflax

This fibrous material comes from Germany's Ore Mountains and is made from chopped **flax straw.** Depending on its area of use, the fibers are cleaned and cut to a defined length. For the nonwoven fabrics industry and applications such as **geotextiles,** interior lining of automobiles, or **impact noise insulation,** the fiber length is between 50 and 80 mm. Short flax fibers are used for the reinforcement of PP and PLA injection moldings. In the construction industry, flax fibers are also used for reinforcing clay plaster.

Bulrush fibers

Bulrushes consist of long, tear-proof fibers and a spongy tissue, a combination not found in other plants. The fibers can be used as a basis for the production of components, for which 90% less energy is required compared with the defibration of wood. Developers at Naporo have managed to activate the **waxes** and oils that protect **bog plants** from moistureas a bonding agent. Moreover, the company has worked with K3P Innovations to develop a **protein glue** noted for its excellent adhesion and extreme resistance to heat. The fiber moldings have a very smooth surface so there is no need for time-consuming smoothing processes or interim varnishing.

Rye and wheat straw

Over the last few years, a series of research projects have looked into the use of rye and wheat straw for fiber reinforcement. Good results were achieved, most significantly, in the production of wood replacement materials like wood plastic composites (**WPC**). The production of composites incorporating polypropylene (PP) was also successful. In the construction industry, research is currently being carried out into the use of straw in connection with cement as a lightweight sheet material. Some sheet materials have now been launched on the market as wood substitutes.

Maize fibers

The long and silky-smooth **maize fibers,** found at the tip of the **corn cob,** store heat very well, are moisture-resistant, and are therefore traditionally used as a cushion filler or in mattresses. The fibers have an antibacterial effect and are naturally flame-resistant. This means they are of interest for other design applications too.

Alfa grass

This fibrous material is cultivated predominantly on the Iberian Peninsula and in certain areas of North Africa. It is well known in the paper industry, for example, where its excellent robustness means it is used to produce high-quality paper that is soft yet opaque. **Alfa grass** can also be woven into slippers, ropes, and bags.

Agave fiber

As a natural fiber, **agave fiber** boasts unusually high resistance to acids and alkaline fluids and remains stable and extremely tough even when subject to heat. This is why the yellow fibers are popular for use in brushes of all kinds, where their good elasticity and water absorption are advantageous. They are obtained from the leaves of specific **agave plants** in the highlands of Mexico.

"Modular thatch panel"
(Design: Ratia Rabemananoro)

Tururi seedpod fibers

In the spring of 2012, a composite material was developed using the **Tururi** seedpod and **natural latex,** which is suitable for environmentally friendly production of biodegradable shoes in the Amazonas region. The self-regenerating seedpods of the Tururi are gathered on the river island of Maraj in a controlled harvest from wild growth. Since palms and rubber trees produce more oxygen over their life cycle than the amount of carbon dioxide resulting from the production and transport of the material, the ecological balance of the semifinished product is positive. This substitute for petroleum-based, nonwoven fabrics is not only suitable for shoe production, but is also used as upholstery material, for sports equipment and fashion accessories, as well as for reinforcing casing surfaces and in detailed solutions for automobile interiors. The prototype "Sęlva 01" was awarded a prize in Rio de Janeiro as part of the UN Conference on Sustainable Development (Rio+20).

BananaPlac installed in the bathroom
(Source: Barkcloth)

Biodegradable "Selva" shoe
(Source: Barkcloth)

Construction materials made of bamboo strips (Source: Conbam)

Properties low weight // sound insulation properties // promotes a healthy atmosphere // formaldehyde-free // suitable for fire protection

Sustainability aspects based on organic waste // biodegradable // good heat insulation properties

STRAW MATERIALS

The cultivation of cereals, pulses, and oil plants results in huge quantities of straw as a waste material. Part of this is reused in agriculture as a nutrient or is used for animal husbandry. In Germany, up to 30% of the available straw can be used for energy production or as a material, though in other countries the proportion is significantly higher. Traditionally, straw is processed in clay construction, made into baskets or hats, and used in the construction industry in bale form as an insulating material (thermal conductivity coefficient: 0.038 watts per meter Kelvin). New materials have been developed recently, in which straw is used as a biomass.

Modular thatch panel

One example is a modular **straw panel** by the French designer Ratia Rabemananoro, which was awarded one of the top prizes at the adream sustainability competition at the end of 2012. In it, the designer uses the straw fibers both for insulation purposes and as a construction material for the external wall. The straw panels are applied to wooden frames according to the principle of **drywall construction.** On the inside, the straw panels consist of straw cob and an earth insulation that is covered with a sheet of OSB. The outer section is clipped in place from the outside using metal pegs and consists of another OSB sheet and a straw covering.

Strawtec wall system

Strawtec is a wall system made of highly compressed **crop straw** which, along with numerous positive aspects of sustainability, also boasts outstanding properties of structural physics. The natural crop straw is pressed without needing a bonding agent and is laminated with an outer layer of recycled paper. Strawtec panels are therefore 100% biodegradable and have a positive effect on promoting a healthy interior. The wall system offers **sound protection** for up to 55 decibels and **fire protection** up to F90. Since no specific framework is required for its assembly, architects can assume costs comparable to those of conventional **drywall construction systems.**

istraw

These straw construction panels consist of a core made of compressed straw, which is condensed in an extruder press process without the addition of a bonding agent. The **lignins** contained within the straw ensure stalks remain bonded together in the long term. Finally, the compressed straw is covered with recycled cardboard, creating a stable and simple-to-process construction panel with sound-insulating properties.

Kirei Wheatboard

This 12.7 or 19.1 mm thick panel material from the USA makes use of straw as a biomass, replacing **wooden fibers** as a typical base substance for wood-based materials. It is therefore free from formaldehydes. There are no negative effects on the environment from the bonding agent used. Its practical properties are comparable with those of medium-density fiberboard (MDF).

Novofibre

The sheet material Novofibre also consists of 100% wheat straw. The organic waste material is processed with a formaldehyde-free bonding agent under heat and pressure. The material is used predominantly for interior applications and is particularly suitable for interior construction and the furniture industry. The sound insulation properties of the material are particularly striking. Given its aesthetic similarity to oriented strand board (OSB), Novofibre has appeared on the market as **Oriented Structural Straw Board** (OSSB).

Straw paper

In a relatively new area of application, **straw fibers** are processed to make paper. This development comes predominantly from Asia, where **rice straw** is used to make particularly lightweight paper. What is remarkable is its coarse-fibered structure, which is particularly advantageous in the area of design.

istraw straw construction panels (Source: istraw)

Structure of the straw panel (Design: Ratia Rabemananoro)

OSSB sheet (Source: Novofibre)

BULRUSH MATERIALS

Bulrush reeds (Latin typha) grow as wild plants in wetlands in all climatic zones. Their cultivation achieves a particularly high yield equating to four times that of wood. The natural supply would be sufficient to satisfy 10 times the European market for natural insulation. The spongy tissue of the leaves, which measure up to 4 m long, offers as good as limitless potential for heat insulation and lightweight construction.

Properties low weight // 4 m long leaves // spongy structure for thermal insulation // diffusion-open structure ensures a good room climate // good fire resistance // pest- and rot-resistant

Sustainability aspects energy-saving due to lightweight construction // excellent thermal insulation // capture of carbon dioxide // high biomass yield

MATERIAL CONCEPT AND PROPERTIES

The insulation effect (heat conductivity: 0.04 watts per meter Kelvin) is a product of the **air pockets** in the spongy structure which are not found in other plants used for insulation like flax, hemp, or straw. Since this marsh plant gets the nutrients it needs to grow entirely from the water at its base, no fertilization is necessary. Drained swampland is ideally suited to the cultivation of these plants. Another advantage for their use in insulation panels and for blow-in insulation of a structure is their extremely low weight (specific density of 65 kg per cubic meter). The pressure-resistant hemp fiber insulating board was awarded the Austrian environment seal in 2013.

The insulation material offers not only thermal insulation, but also improved sound insulation. The diffusion-open structure has a positive effect on the **room climate,** since the moisture is transported effectively from the inside outwards. Another advantage of the insulation material is its notable fire-resistant properties. In case of fire, the material chars on the outer surface, but further damage is avoided in a natural way. Given their origins as marsh plants, bulrush-based insulation materials are pest- and rot-resistant.

APPLICATION

The pressure-resistant insulation material is marketed by Naporo predominantly for thermal insulation of facades, attics, and roofs. It is also now available as a **blow-in insulating material** for wall cavities or inter-rafter insulation. Alongside the entirely recyclable insulating material, other bulrush-based products are now available on the market, including the Q-Fight lightweight construction panel with high stability and a particularly low weight. In future, it is hoped that the bulrush plant may be used as a biomass for foodstuffs and pharmaceuticals.

Bulrush reed
(Source: Naporo)

Spongy structure of the bulrush plant
(Source: Naporo)

Compostable insulation panels made of bulrush reeds with bonding fibers made using starch (Source: Naporo, photo: Diana Drewes)

Bulrush-based lightweight panel with high compressive strength (Source: Naporo)

Lightweight construction panels made of a bulrush-based spongy structure
(Source: Naporo, photo: Diana Drewes)

Properties more soft and porous than wood // lower weight // strong rigidity // unusual texture

Sustainability aspects capture of carbon dioxide // material based on waste products // high biomass yield

SORGHUM MATERIALS

While in Europe, flax and hemp are the most frequently cultivated fast-growing plants with a high biomass yield, in Central America, Southeast Asia and sub-Saharan Africa their equivalent is sorghum. Its importance as an energy plant for gas and electricity production has increased over the last few years. In the USA, sweet sorghum is most commonly used for the production of bioethanol.

MATERIAL CONCEPT AND PROPERTIES

Around 20% of global sorghum production now takes place in the United States. It therefore stands to reason that this fibrous material should also be used for material production. The structure of the sorghum plant is similar to that of **maize.** One example of a **sorghum sheet material** is **Kirei Board,** which is produced in the USA using the coarse structures of the fibers and stems of the plant. The manufacturing process ensures that, despite their low weight, the sheets are extremely rigid. Compared with wood, the material is softer and more porous.

APPLICATION

With its unusual texture, Kirei Board is most frequently used for furniture construction, interior design, wall cladding, or for decorative products. Due to its soft structure, however, it is not suitable for use on well-used floors or for external applications. Kirei Board can be processed using all the conventional procedures. Furthermore, it is compatible with all commonly found varnishes and paints and is available in thicknesses of 6, 10, 20 and 30 mm.

Texture of Kirei Board
(Source: Kirei Board USA)

Sorghum cultivation in Germany
(Source: KWS Saat AG)

The water hyacinth is a tropical plant species whose rapid spread in some areas of the world is a growing cause for concern. At the same time, this plant offers considerable potential with regard to applications in the packaging, paper, and furniture industries.

Properties high tear-resistance in the direction of the fibers // translucent fibers // can be processed as pulp // low-maintenance cultivation

Sustainability aspects quickly growing biomass // purification of bodies of water polluted by heavy metals // absorption of oil films

WATER HYACINTH FIBERS

MATERIAL CONCEPT AND PROPERTIES

The fibrous material of the water hyacinth has sufficient tear resistance in the direction of the fibers, but not as much as alternatives made of **bamboo** or **rattan.** Due to its cost advantage and low-maintenance cultivation, however, over the last few years it has increasingly attracted attention as a source of fibers. An unusual aspect is the water-based plant's rapid growth, which makes it particularly appealing for energy generation in biogas plants. Water hyacinths produce three or four new offshoots every day and reach full size after four months. The plant is able to store **heavy metals** within its structure and to absorb **oil films** from the surface of the water.

APPLICATION

The dried stems and strands of the water hyacinth can be used in **wattling,** as a filler in composite materials, or as an organic packaging material. Water hyacinth fibers have also captured the imagination of the paper industry. The translucent nature of the fibers offers potential for lighting and possibilities for the implementation of room installations. The strands of the water hyacinth can also be unraveled and made into **pulp** to be pressed into moldings and subsequently dried.

Furniture made of water hyacinth fibers (Source: Projektwerkstatt Potsdam)

Water hyacinths (Source: Projektwerkstatt Potsdam)

Water hyacinth fibers being processed (Source: Projektwerkstatt Potsdam)

Even hundreds of years ago, stinging nettles were known as a fiber source for fabric production. With the increasing importance of natural fibers for the textile industry, they are regaining the attention of producers. Their cultivation does not require any fertilizer or artificial watering.

Properties long, tear-proof fibers // free from chemicals // breathable // tolerable for allergy sufferers // nettle fabric as an alternative to cotton

Sustainability aspects low water consumption for cultivation // no fertilizer necessary // biodegradable

STINGING NETTLE FIBERS

MATERIAL CONCEPT AND PROPERTIES

Long, tear-resistant fibers are obtained from the stems of stinging nettles, which can be made into particularly tear-resistant materials. The natural color of the fibers ranges from cream to brown. **Nettle fabrics** are free of chemicals, breathable, and suitable for allergy sufferers and, most importantly, shirts made of natural fibers are extremely comfortable to wear. The fabrics are soft and shiny and represent a regional alternative to cotton: beyond a minimum area of 10,000 hectares, there is a cost advantage over **cotton fibers** from India. In addition, nettle fabrics are compostable.

APPLICATION

The main area of application for stinging nettle fibers is in textiles, where they are particularly suited to jeans, bed linen, and shirts. With its absorbent properties, this fiber material is useful for other applications too. In addition to their use in clothing, nettles can be pressed into nonwoven fabrics for the interior lining of vehicles. Very recently they have been used in packaging as well. The **stinging nettle root** can also be used to dye textiles.

VVIO

This composite material by designers Eva Marguerre and Marcel Besau consists of common **stinging nettle fibers** and a bioplastic made using **potato starch,** and was developed for use in small pieces of furniture. The **bast fibers** obtained from the nettle are initially processed into large sheets of nonwoven fabric and coated with the plastic. After heat treatment, the material boasts high stability and can be colored with natural pigments. Since only natural resources are used, the fiber composite material is biodegradable.

Stinging nettles in the wild

Stinging nettle fibers dyed with natural substances (Source: Eva Marguerre and Marcel Besau)

Nettle fiber bioplastic composite (Design/source: Eva Marguerre and Marcel Besau)

VVIO stinging nettle fiber bioplastic composite with gel-like layer of potato starch (Design/source: Eva Marguerre and Marcel Besau)

Flax and linen fibers are increasingly used for reinforcement of composite materials due to their good mechanical properties and low price.

Properties very good tensile strength // medically harmless // low weight // good CO_2 balance // problem of water absorption // densities: 400–2,000 g/m²

Sustainability aspects based on renewable raw materials // biodegradable // low energy requirements in production

FLAX FIBER COMPOSITES

MATERIAL CONCEPT AND PROPERTIES

Linen fibers are obtained from the stems of the **flax plant.** They are highly tear-resistant and rigid, and are medically safe. Used in composites, they boast very good tensile strength compared with those made using **fiberglass,** and are particularly noteworthy for their low weight and good CO_2 balance. When using them, however, it is essential to pay attention to the high water absorption of the natural fiber composite, since this can have negative effects on the bonding strength of the fibers in the matrix. Once the flax compounds have dried out they regain their original mechanical properties. The choice of fiber material, matrix plastic, mix ratio, and bonding process can be adjusted to determine the material's qualities, spanning a broad range. Depending on the intended purpose, the material's density is between 400 and 2,000 grams per square meter.

APPLICATION

Over the last few years, flax fiber composites have increasingly been used in the automobile industry and in the area of electronics. They are usually made into parts for internal door cladding, fittings, or casing components for electrical appliances through compression molding. **Short fibers** can also be processed through injection molding or extrusion. Two thirds of the natural fibers used in the automobile industry are flax fibers.

PRODUCTS

ampliTex®

Under the brand name ampliTex®, Swiss company Bcomp has developed new **flax fiber composites** that boast greater rigidity, flexural stiffness, and better insulation properties than traditional materials. Due to the particular mode of fiber integration, the flexural stiffness of sheets and pipes can be improved by two and a half times with no fiber reinforcement and with an increase in weight of just 5–20%. The so-called **power ribs** can therefore compete with **carbon fiber reinforcements** and are already being used in the sports industry – for example a ski made of this material weighs around 30% less.

Fiber structure for the production of so-called power ribs (Source: Bcomp)

Flax fibers as a reinforcement material (Source: Bcomp)

Biotex fiber materials (Source: Composites Evolution)

bTubes

These tube profiles made of flax fibers and resins boast particularly high rigidity and have three times the insulation capacity of comparable carbon fiber tubes. This means bTubes with diameters of 18 and 22 mm have ideal properties for use in sport and trekking applications.

Biotex

British manufacturer **Composites Evolution** produces a series of fiber materials (flax, flax/PLA, flax/PP) for lightweight construction biocomposites under the brand Biotex. The fibers, strips, and sheet textiles can be processed using various finishing technologies such as thermoforming, wet coiling, pultrusion, or injection molding. Biotex materials are generally used in interior applications in automobile construction, for furniture design, or in the building trade. However, they are also suitable for outdoor use as they possess antibacterial and moisture-proof properties.

e2e Materials

An offshoot of Cornell University, the company **e2e Materials** specializes in the development and production of 3D biocomposites for furniture construction and interior design. For the production of biodegradable sheets of chipboard and MDF, flax fibers in particular are combined with a **soya protein adhesive** and compacted using pressure and heat. The process uses just 19% of the energy of conventional production methods. **Jute** and **kenaf fibers** are also used.

Lineo flax fibers

Manufacturer Lineo specializes in the integration of flax fibers in sports items such as tennis rackets, bicycle wheels, golf clubs, and crash helmets. These applications make optimum use of the flax fibers' properties for lightweight structures, vibration damping, and impact absorption. The flax fibers are generally processed with thermosetting resins.

3D biocomposite made of flax fibers (Source: e2e Materials)

bTubes have particularly good rigidity and insulation properties (Source: Bcomp)

Relative flexural stiffness of flax fiber composites (Source: Bcomp)

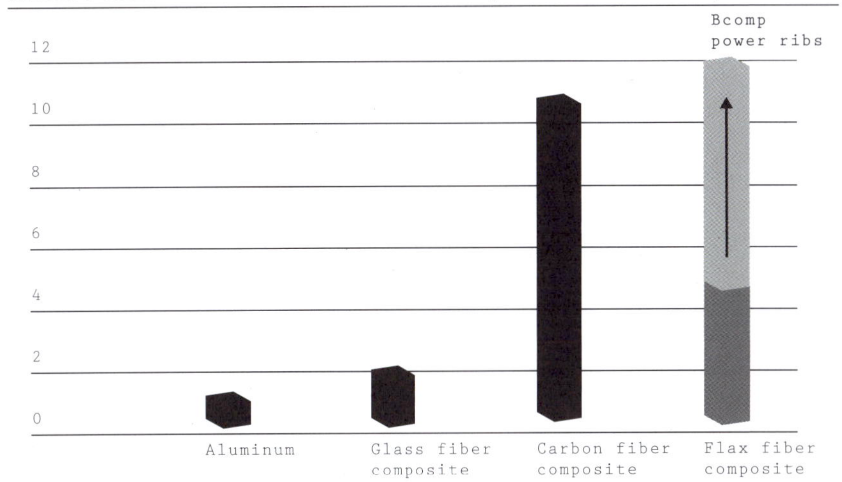

Properties wood substitute // stores heat // low weight // hardwearing // basis for art production

Sustainability aspects no negative consequences for food prices // organic waste products save finite resources

UNUSUAL ORGANIC PARTICLES

The growing significance of natural sources of raw materials for use in architecture and product development has meant unusual organic waste materials are an increasingly frequent topic of discussion. Most appealing are waste products that have no other uses and do not have a negative impact on food prices.

MATERIALS

Tea powder

Chinese artist Ai Weiwei's **Tea House** from 2009 has become particularly well known in connection with this. It consists of 378 cubes and 54 prisms, which were produced from used **tea powder.** The installation not only demonstrated the possible uses of organic particles in architecture, but also highlighted the potential of natural materials to appeal to several of our senses at once. The Tea House gives off the scent of the raw material in the vicinity of the emperor's throne in the East Asia art collection in Berlin-Dahlem, so symbolizing the cultural treasures from the glory days of the Chinese empire.

Rice husks

A new weather-resistant and waterproof material is being marketed under the name Resysta. It is made of 60% **rice husks,** together with 22% rock salt and 18% mineral oil. This structure gives the material a particular resistance to moisture. It is therefore suitable as a substitute for **tropical wood** and is virtually identical to it in appearance. It has initially been used for outdoor furniture, floor coverings, and ships' decks. Outstanding resistance to mold and fungal growth means Resysta is suitable for high-tension house facades and wet areas.

Cherry stones

Cherry stones store heat for a particularly long time compared with other materials and then release it very gradually. They are therefore well suited to use as fillers for heat cushions for the relief of stomach ache. Since cherry stones only release the heat very slowly, there is no risk of burns. As a rule, the stones are heated in the oven at temperatures of around 100 °C. Contact with an open flame should be avoided. Cherry stone cushions are also suitable for cooling purposes.

Apricot and peach stone particles

The stones of fruits like apricot or peach are particularly hard and resilient. BioGranulats® are fragments of these **fruit stones,** and are suitable as a natural and aesthetically pleasing alternative to chippings or asphalt for driveways to homes and pathways. The manufacturer predicts a life cycle for the particles of 15 years and calculates 30 kg or 50 liters for an area of one square meter at a depth of 50 mm.

Driveway with apricot or peach stone particles (Source: BioGranulats)

Weatherproof wood substitute made of rice husks (Source: Resysta, photo: Diana Drewes)

Cherry stones as a filler for heating and cooling cushions

Biomaterial made of wood shavings (Source: Organoid Technologies, photo: Marion Luttenberger)

Egg shell composite material (Source: Ulrike Böttcher)

Crystallized material made using egg shells (Source: Ulrike Böttcher)

Nut and stone fruit shells

Hazelnuts, pistachios, peanuts, or macadamia nuts: once the core has been enjoyed, all these delicacies leave behind a piece of extremely strong organic waste that lends itself to use as a particle component in composite materials. **Macadamia** nuts have the hardest shells.

Wood shavings

With their thermal insulation properties, wood and wood shavings have been used in a whole range of solutions since the 1990s. One of the best known insulating materials is HOIZ, which is refined with untreated **fresh whey** with a few percent of **soda leaching additive.** Impregnation with whey reduces the fire risk, thus achieving fire protection class E. One cubic meter of in-built **wood shaving insulation** stores around 50 kg of carbon dioxide. When made into sheet materials, wood chips and shavings also offer aesthetic qualities.

Egg shells

Egg shells are easily broken down and are 100% biodegradable. By combining them with a bonding agent, designer Ulrike Böttcher has developed a pourable material which, when hardening, behaves in a similar way to mortar and can be used as wall plaster in indoor spaces. The material is lightweight and the varied distribution of the components means the egg shell quota can still be recognized. In addition, the high proportion of lime provides for good moisture distribution, promoting a positive room climate.

With the help of a biological solvent, the designer also discovered that the surface can be modified in a specific way. A crystallization process is triggered which considerably changes the surface of the material and creates a white, needle-like structure. This process can also be created using coatings on paper, ceramic, and textiles.

Mussel shells

The shells of mussels are made from **lime.** Their use in the area of crafts and jewelry is no doubt a familiar sight from summer holiday resorts. Less conventional, however, is the use of finely ground mussel lime in the construction industry as well as in art. Artist Giovanni Manfredini uses **mussel flour** and resin as a base for his painting. After application on the paint surface, the mixture is dried and then colored black with a Bunsen burner. The dark background enables the artist to print body parts and images with very strong black and white contrasts, achieving an X-ray-like appearance.

Diatomite

The porous **diatomite** powder consists mainly of **silicon dioxide** and is obtained from **diatom shells.** Diatomite is very low in weight, extremely resilient, and prevents the dispersion of dyestuffs. It is therefore prized predominantly as an additive for construction materials, paper, and plastics. At the Fraunhofer IWMH in Halle, researchers are currently working on the development of bio-based tiles made of diatomite, natural fibers, and linseed oil epoxy. They can be manufactured in a much more energy-efficient way than conventional solutions. The format and color can be tailored to individual requirements, and adding appropriate pigments makes it possible to create tiles with phosphorescent and heat or water-sensitive effects.

Biotiles made using diatomite
(Source: Fraunhofer IWMH)

Work by the artist Giovanni Manfredini
(Source: Giovanni Manfredini)

HORN

Properties extremely hard // very hard-wearing // absorbs moisture // sensitive to vibrations // no allergic reactions

Sustainability aspects based on organic waste // biodegradable

Animal horn is made from keratin. It is used by designers, particularly for high-quality spectacle frames, as a natural material and a replacement for plastic.

MATERIAL CONCEPT AND PROPERTIES

The relevant horn material comes from horned animals such as cattle, sheep, goats, buffalo, or yaks and is a waste material generated at slaughterhouses. In terms of hardness and wear resistance, the material is superior to most woods and, more importantly, boasts excellent compressive strength. Depending on the type of animal it comes from, the horn color varies between blackish-brown and yellowy-beige. This can, however, be influenced by the type of processing. Like other natural products, horn also absorbs moisture from the environment, which has an effect on its properties. If the horn comes from wet regions, care must be taken that it is not dried out too quickly, which carries the risk of cracks forming. One important factor concerning contact with human skin is that horn does not provoke any allergic reactions. What's more, it is taste-neutral, and as such is suitable for high-quality cutlery.

APPLICATION AND PROCESSING

Horn has been recognized as a practical natural product since the Middle Ages. Typical products it has been used for include spoons, knife handles, and hair pins. Using combs made of horn avoids irritation of the scalp and helps to prevent dandruff, and cutlery made of horn is ideal for savoring delicate foodstuffs such as caviar.

The material can be worked using typical processes with saws, files, or drills and can be polished with pumice powder or raw leather. When heated to temperatures between 160 and 190 °C, horn can be bent. For thinner pieces, a hot air dryer is generally sufficient to allow the required distortion. To maintain the condition of horn surfaces, vegetable oils or Vaseline can be rubbed into them from time to time.

London-based designer Tom Davies has developed a production process to manufacture high-quality spectacle frames out of buffalo horn. The horn is first cut into pieces, then stored for six to seven months in water and under pressure, then subsequently polished. To ensure the material cannot regain its original curved form, it is glued to sheets with the opposite curve using a **keratin adhesive** obtained from the hooves of the buffalo. Processing of this delicate material takes place at a specific level of humidity and at temperatures between 15 and 24 °C. Since **buffalo horn** reacts sensitively to vibrations, a **laser** is used for the individual cut rather than a milling cutter. The British designer uses this same process to produce **horn buttons** in place of plastic ones.

Manufacture of spectacle frames made of buffalo horn (Design: Tom Davies, TD Tom Davies Ltd., S. Jakub)

Spectacle frames made of buffalo horn (Design: Tom Davies, TD Tom Davies Ltd., S. Jakub)

Properties moisture-proof or biodegradable on contact with water // fertilizer // protects against UV radiation // odor-neutralizing

Sustainability aspects based on organic waste // biodegradable

COFFEE GROUND MATERIALS

With their beakers and lampshades made using the waste from espresso machines, designers Julian Lechner and Raúl Laurí have demonstrated that coffee grounds as an organic waste material can also be used in product development. In other designs, designers use the nutrients in coffee grounds for the creation of flower pots or urns for animal carcasses. Coffee grounds also offer potential for dying wooden surfaces.

MATERIALS AND PRODUCTS

Coffee ground moldings

For the production of moldings made of coffee grounds, the waste material is combined with a bonding agent and pressed into a mold under the influence of heat. The choice of bonding agent depends on the purpose of the molding. With the addition of caramelized sugar, for example, it is possible to produce mugs that gradually dissolve as coffee is drunk and give the coffee an additional aroma. This option allows the user to enjoy around 20 cups of coffee. After use, the cup can be disposed of with the organic waste. A dishwasher-proof variant, which would also be suitable for series production, is being made by Julian Lechner using a bioresin as a bonding agent. Designer Raúl Laurí has followed a similar route. In addition to drinking vessels, he has also used coffee grounds to produce lampshades and tiles.

It is widely known that coffee grounds have a positive effect on plant growth. The nutrients they contain enrich the soil and also beneficial for house plants or the compost heap in the garden. These advantages were exploited by designer Sanam Viseux in 2012 for the development of compostable flower pots, which contain a natural fiber mixture and wax in addition to the coffee grounds, and biodegrade in earth.

Çurface coffee ground wood

British company Re-worked has developed a composite material made of coffee grounds, recycled plastic, and wood fibers. The material is combined with used wood to produce furniture. Alternatively, the coffee ground wood can be used for the casing of a coffee machine.

S.Café® coffee ground fibers

In Taiwan, a company called SINGTEX has registered a patent for a technology that produces fiber material using recycled PVC bottles and coffee grounds, and does so in a low-energy, toxin-free way. The S.Café yarn can be worked to produce cloth or can be used directly in the clothing industry. Coffee ground fibers protect against the impact of UV radiation, neutralize odors, and dry very quickly. This makes the fibers particularly appealing for use in sports clothing. Liverpool Football Club has already used the coffee ground fibers in their shirts.

Surface made of bonded coffee grounds
(Design: Raúl Laurí)

Flower pots made of coffee grounds
(Design: Sanam Viseux)

Lampshades made of coffee grounds
(Design: Raúl Laurí)

Coffee cup with bonding agent made of caramelized sugar (Design: Julian Lechner)

Coffee machine with surface made of coffee grounds
(Source: Adam Fairweather, Re-worked)

Table made of coffee grounds
(Design: Julian Lechner)

Bowls made of coffee grounds
(Design: Raúl Laurí)

Functioning mechanism of S.Café coffee-ground fibers (Source: SINGTEX)

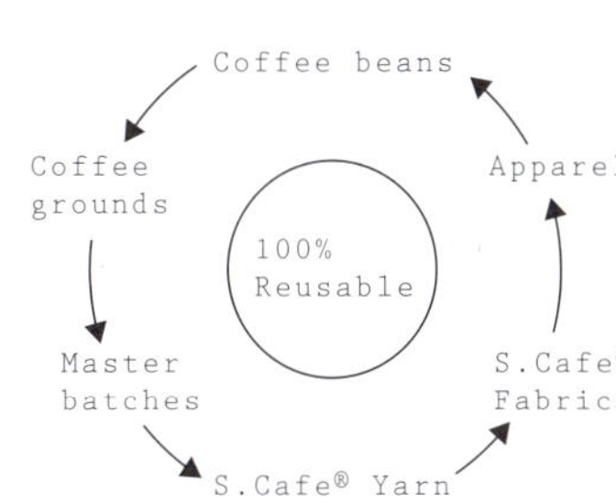

Fish scales are a waste material from the fishing industry, which produces them in large quantities. During a project with school students in a township of Cape Town, designer Erik de Laurens made an astounding discovery, and in doing so developed a new material.

Properties moisture and heat-sensitive // thick-walled components

Sustainability aspects based on organic waste // no chemical additives // biodegradable

FISH SCALE PLASTIC

MATERIAL CONCEPT AND PROPERTIES

Erik de Laurens was able to use fish scales to develop a material that can replace plastic for the manufacture of moldings like spectacle frames and beakers when heat and pressure are applied, with no need for a bonding agent. Thus a new biomaterial based on organic waste has joined the series of unusual discoveries over the last few years. The formation of the plastic is down to the collagen contained in the fish scales, which acts like a natural adhesive under the influence of heat. The material is 100% biodegradable.

APPLICATION

So far, Erik de Laurens has used the new material to create standard objects such as spectacle frames, drinking vessels, and inlays for furniture surfaces. Since its mechanical properties, particularly its rigidity and resistance to heat, have not yet been extensively researched, the designer is concentrating on thick-walled components. He is currently seeking a financial backer in order to ready the material for industrial production.

Furniture inlays made of fish scale plastic (Source: Erik de Laurens)

Spectacle frame made of fish scale plastic (Source: Erik de Laurens)

ALGINATE

One algae-based material that is known particularly in the world of dentistry is the exceedingly elastic impression material known as alginate.

Properties rubber-like consistency // high level of precision when used to make impressions // short bonding time // tear-resistant // significant shrinkage // skin-friendly // biocompatible

Sustainability aspects based on renewable raw materials // no competition with foodstuffs // optimization of energy storage possible

MATERIAL CONCEPT AND PROPERTIES

Alginate is obtained from extracts of **brown algae** and **seaweed** and can be used for extremely accurate, pore-precise impressions of areas of the body with a short bonding time of just a few minutes. The hardened mass has a soft, rubber-like consistency and is extremely tear-resistant. When used for impressions, the alginate will show greater shrinkage than **silicon.** Alginate is skin-friendly, physiologically harmless and biocompatible, and permits the incorporation of human cell tissue.

APPLICATION

Alginate is used as an **impression material** with a silicon-like consistency for art, in particular, and in the semiprofessional field for life-casting (although not for undercut areas). In dentistry, it is used for making impressions of sets of teeth. **Calcium alginate** compresses are used as **wound dressings** and prevent the bandage from sticking to the wound. In 2011, scientists from Atlanta successfully used alginate for the production of **lithium-ion batteries** made of fine silicon powder. Compared with conventional graphite electrodes, a storage capacity eight times higher could be demonstrated. Previous attempts to produce batteries using silicon failed due to the significant volume change during the charging process. The swelling is prevented by the algae putty. The pores that form during production reach optimum dimensions in terms of size and quantity for particle migration and storage.

Alginate impression material
(Photo: Diana Drewes)

BAGASSE

Properties contains lignin // formaldehyde-free flat laminate possible

Sustainability aspects based on renewable raw materials // recyclable // waste product from sugar production

Bagasse is produced as a by-product during the extraction of sugar from sugar cane after the syrup has been squeezed out. The extraction of 10 tons of sugar will produce around 3 tons of the waste material, which in many cases is simply incinerated. Bagasse consists of around 20% lignin in addition to its proportion of cellulose. The hemicellulose also contains the polysaccharide Dylan, which can be used for the production of bioplastics and biomaterials in bio-refineries.

zhè – furniture made of bagasse material
(Design: Chen Wei-che, Chung Yo-Hsun)

Eco HPL

Using bagasse as a basis, laminate expert Dekodur has produced the world's first formaldehyde- and phenol-free flat laminate, named Eco HPL, which consists of up to 100% biological waste materials. The material's formaldehyde emission is given as < 0.01 parts per million (EN 717/1), which makes the material particularly valuable for use in interior spaces.

Barkcloth Dekowood laminate
(Source: Barkcloth)

RE-Y-Stone

This material consists of recycled core and wallpapers and furan resin that is obtained from bagasse as a waste material from sugar production. Its hard-wearing surface is long-lasting, mechanically highly resilient, and dimensionally stable. With these properties, the recyclable material can be used both in flooring and in furniture and interior design.

Dekowood Barkcloth

Barktex, the manufacturer of Barkcloth, has managed to use Eco HPL to launch an entirely biodegradable flat laminate, which combines the look of the tree bark textile from Uganda with the environmental compatibility of the bagasse-based HPL.

RE-Y-Stone surface
(Source: Resopal)

Up to the 18th century, candles were made almost exclusively from beeswax, which still achieves the highest qualities today. During the course of industrialization, production was shifted almost entirely to paraffin, which results from the lube oil distillation of crude oil. Against a background of rising raw material prices and the future bottleneck of crude oil, over the last few years there has been a return to industrial production of candles made using renewable raw materials.

Properties low-soot burning properties // minimal dripping // velvety-matt surface // long shelf life

Sustainability aspects based on a renewable raw material // reduction in greenhouse gas emissions // regional production possible

Conventional
candles

RAPESEED CANDLES

MATERIAL CONCEPT AND PROPERTIES

For the production of **pillar candles,** the suitability of **rapeseed oil** from agriculture has qualified for fully automated production using typical processes like granulation, pressing, extrusion, or pulling. An initial solution based on rapeseed oil shows good, low-soot burning properties. Another striking feature of this bio-based material is its tendency not to drip.

APPLICATION

Rapeseed candles are characterized by a velvety-matt surface and an exceptionally long shelf life. Their high-quality appearance makes them particularly suitable for classic functions in the field of decoration.

Properties flexibility and color fastness is comparable to leather tanned using mineral salts // tight-grained // dimensionally stable // breathable // skin-compatible

Sustainability aspects leather treatment using renewable raw materials // no use of environmentally harmful chrome salts and metals // no allergic reactions // biodegradable

NATURALLY TANNED LEATHER

Leather has been the standard material for shoes, bags, belts, and hats for centuries. This flexible, yet resilient material is made using animal skins that become durable in a chemical tanning process. This involves three steps: first the skins are preserved through drying or salting and prepared for storage or transportation. The subsequent soaking removes particles of dirt, blood, and preservatives. Heat and moisture loosen the hairs, which are removed through the addition of lime and sulfur compounds. The actual tanning process takes place using plant-based, mineral, or fatty tannins.

Due to the mechanisms of industrial production, natural tannins have generally been replaced by cheaper chromium salts, which have very negative effects on the environment. This is because the sludge produced during the leather production contains heavy metal deposits. What's more, when old leather is disposed of through incineration, highly toxic chrome may be released. Consequently, there are increasing efforts to replace environmentally harmful tannins with natural ones.

ALTERNATIVE TANNING PROCESSES

Olive leather

One example is the **wet-green® technology,** which was launched in the summer of 2012 by N-Zyme BioTec GmbH. Using **olive leaf extract tannins,** obtained from waste products from **olive oil production,** the company produces a natural tannin for the production of low-toxin, semifinished leather. The product results in a tanning process superior to most other well-known techniques in terms of simplicity and environmental responsibility. The olive leaf extract tannin is free from **chrome,** nickel, and other metals, and its production helps to reduce the environmental burden in olive-producing regions. The resulting tanning waste and effluent are nontoxic and easier to dispose of or to recycle (for example in **biogas plants**). **Olive leather** is very tight-grained and extremely dimensionally stable. Its colors can be compared to leather tanned using mineral salts. The **bioleather** is suitable for use in the automobile, furniture, and shoe industries.

Rhubarb leather

Another possibility for environmentally friendly leather tanning has been developed by scientists at the University of Anhalt over the last few years. They have succeeded in using extracts from the root of the **rhubarb plant** for a tanning process that is free of toxins, heavy metals, and **chrome salts.** As a result, the rhubarb leather is particularly breathable and skin-compatible. It is

Obtaining the olive leaf extract tannins in the laboratory (Source: wet-green GmbH)

Bag made using rhubarb leather (Source: deepmello)

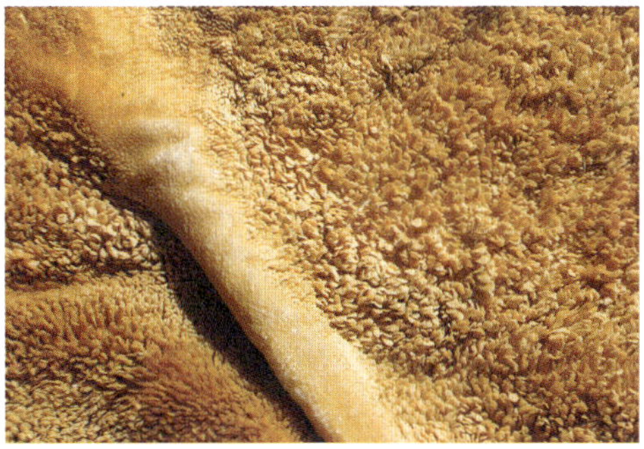

"Rumen Leather" – leather made of cow stomach (Source: Mandy den Elzen)

especially suitable for people who suffer from skin allergies. In 2011, rhubarb leather was awarded the ECARF quality seal by the European Centre for Allergy Research Foundation. It is suitable for the production of leather bags as well as for belts and accessories. Rhubarb leather is easily biodegradable and can be returned to the natural cycle with no harm caused.

ALTERNATIVE SOURCES OF LEATHER

Bovine stomach leather

An unusual, but aesthetically particular base material for leather production is the **cow stomach (rumen).** It features a honeycomb structure, making it particularly interesting for accessories and bags. Dutch designer Mandy den Elzen has made a name for herself with her work using cow stomach material under the name "Rumen Leather". The pieces measure up to 400 x 500 mm and are 3 mm thick.

Fish leather

Fish skin is another unusual source for leather production. However, only small pieces of leather can be produced, so the fish leather can only be used in niche areas. Over the last few years, its use has been recognized for objects and accessories, for example for furniture covers, purses, and lampshades. Since waste products from the fish industry are used, **fish leather** contributes to **species conservation** of rare **reptiles.** Fish species particularly suitable for leather production include eels, carp, salmon, trout, and rays. Its manual production has been acknowledged for some time and is practiced in remote regions such as Alaska and Siberia. Fish leather has also been produced industrially in Bavaria since 2007. The durability of this fish leather is comparable to that of better known leather varieties such as cow and calf leather.

Chicken leg leather

Chicken legs also count as organic waste, and have so far rarely been recycled for high-quality applications. Their use in leather production is rare, but from the point of view of preserving endangered species, it clearly makes sense. This is because chicken leg leather is so similar to **reptile leather** that it could help to spare wild **crocodiles.** What's more, **chicken leg leather** is a pure by-product of food production.

Chicken leg leather
(Source: PiraColor)

Object made of fish leather
(Design: Erik de Laurens)

Palm leather

Under the name "Palmleather," designer Tjeerd Veenhoven has developed a material made of **palm bark,** which can replace normal leather in shoes or bags. Strictly speaking, the material's plant origins mean it is not actually a leather but a leather substitute. To soften up the hard-wearing and tear-resistant fibers, the Dutch designer soaks the bark of the **areca palm** in a special biological solution. The fibers become soft and are then suitable for the production of compostable disposable sandals, for example.

Sandals made of palm leather
(Design: Tjeerd Veenhoven)

Properties edible // can be combined with foodstuffs // sufficiently dimensionally stable

Sustainability aspects biodegradable // no environmentally harmful waste products // based on renewable raw materials

EDIBLE PACKAGING

Packaging represents a large proportion of household waste. Every German produces an average of 72 kg of packaging waste each year. Despite the separation of waste, which has now been practiced for some time, only about half of the materials collected are recycled. The rest is incinerated for energy production or disposed of in other ways. Optimizing the cycle of production, consumption, disposal, and biodegradable packaging materials should help to reduce the amount of waste produced in the future. A new idea for food and drink is the use of outer packaging made of edible materials.

PRODUCTS

Füllett®

Just like the **Cookie Cup** by designer Enrique Luis Sarde, the thin-walled **Füllett** is also baked using natural ingredients. In a patented process, cereal flour, rapeseed oil, water, and salt are transformed into a receptacle for canapés, tapas, or baked dishes. It is authorized as a foodstuff, suitable for consumption in soups and sauces, can be stored in the fridge, and has a taste similar to bread.

Blütezeit

Designer Virginia Binsch has created a wrapping made of fold-out elements for the packaging of cheese. The packaging is supposed to be enjoyed together with the cheese in the form of **crackers** to complement the taste. There are therefore a number of different taste variations.

WikiCell

At the renowned Harvard University, David Edwards has developed a waterproof material made of a bioplastic and natural particles which is up to 100% biodegradable. The developer's inspiration was a fruit whose skin protects the delicate inside from outside influences. While the WikiCell skin is made of **sugarcane waste** (**bagasse**), the actual packaging skin is made from the biopolymer **Chitosan, alginates, algae extracts,** and natural particles of nuts, seeds, and chocolate. The skin can be used to transport liquid or gel-like foodstuffs like ice, sauces, cheese, or drinks. It is pierced with a drinking straw to enable consumption.

Blütezeit bread packaging (Design: Virginia Binsch)

WikiCell – edible packaging inspired by fruit (Source: David Edwards)

Drinking vessels made of pectin and sugarcane (Source: Loliware)

Loliware

Edible beakers from New York-based company Loliware are produced using natural fruit **pectin** and **sugarcane.** They are available in five different flavors and were created by graduates of the Parsons Design School. The drinking vessels represent a sensible alternative to disposable crockery and offer an environmentally friendly alternative for occasions such as large-scale events, because once the drink has been consumed, the packaging can be eaten as a sweet treat.

Vivos®

Under the brand name Vivos, US company MonoSol has developed a **water-soluble film** for the packaging of tea, coffee, or fruit drinks, which can be dissolved in both cold and hot water and consumed with its content. The film is transparent, protects against penetration by oxygen or water vapor and, once dissolved, has no effect on the taste of the food.

Water-soluble film as food packaging (Source: MonoSol)

Drinking vessels made of pectin and sugarcane in five different flavors (Source: Loliware)

EDIBLE DESIGN

Properties edible // sufficient stability // use of waste possible

Sustainability aspects based on renewable raw materials // made of food waste // biodegradable // nontoxic

Along with the development of edible packaging, designers are increasingly daring to produce classic design objects and items of furniture using ingredients from the kitchen. Thus the Hard Candy coffee table has a tabletop, for example, made from 5.6 kg of sugar, 2.1 liters of corn syrup, 1.4 liters of water, and 100 g of edible wax.

PRODUCTS

TOFU chair

Designer Leonardo Talarico aims to appeal to vegetarians in particular with his latest furniture design. This is because it consists entirely of **blocks of tofu** that have been dehydrated. With heat treatment, the designer has fine-tuned the stability properties of the tofu blocks to such an extent that they can be used to make an item of furniture.

BAGUETTE tables

The amount of food that is not consumed but simply thrown away is immense. On average, a family of four will throw away around six full shopping trolleys of food every year. In order to draw attention to the Western world's wasteful approach to food, Polish designers Gosia and Tomek Rygalik have developed a series of tables made of a multitude of French-style **baguettes.**

sugarchair

With the sugarchair, Dutch designer Pieter Brenner has created a chair made entirely of sugar. It is available in a limited edition and is made on demand. The individual selection of color and form in the processing of the approximately 30-kilo **sugar mass** gives rise to unique pieces that differ from one another particularly in terms of their taste.

BITE ME

In addition to using edible substances for furniture and crockery, designers are trying them out in electrical products and lighting. American designer Victor Vetterlein, for example, has developed a light made using biodegradable plastic consisting of plant-based **glycerin** and **agar,** a **gelatinous substance** extracted from the cell walls of a type of **red algae.**

TOFU chair
(Design / source: Leonardo Talarico)

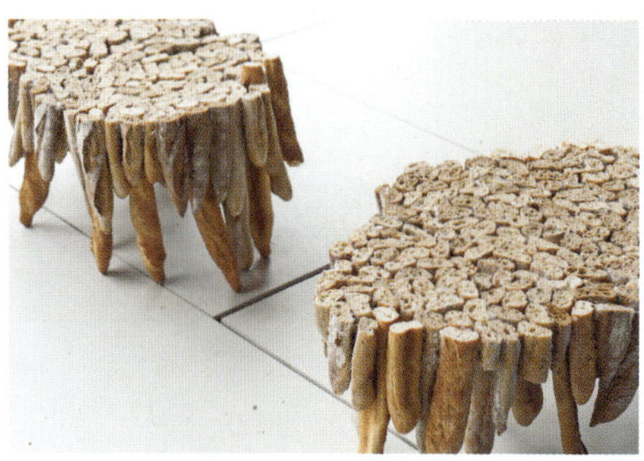

BAGUETTE tables
(Design: Gosia and Tomek Rygalik)

sugarchair
(Design / source: Pieter Brenner)

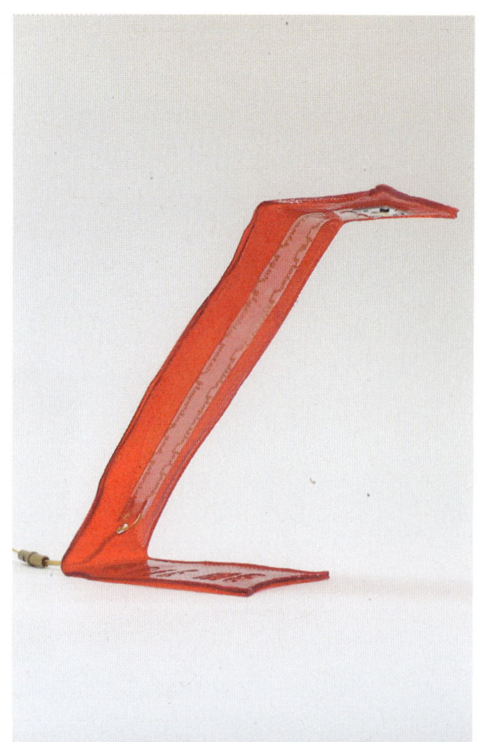

BITE ME – edible light
(Design / source: Victor Vetterlein)

Properties biocompatible // possibility of influencing decomposition speed // implantation in the human body possible

Sustainability aspects made using natural base materials // biodegradable

BIOLOGICAL ELECTRONICS

Electronic scrap and old electrical appliances represent a big challenge when it comes to disposal. Most problematic are the bonds between plastics and metals, the single-origin selection of which is currently only at the rudimentary stage. It is therefore not uncommon to find old appliances from Europe in Africa, where people try to find precious metals among the waste. During this process toxic heavy metals are released, polluting soil and groundwater. Against this background, there are a number of projects looking into biocompatible electronics.

MATERIAL CONCEPT AND PROPERTIES

At the Johannes Kepler University (JKU) in Linz, researchers have developed an organic field-effect transistor based entirely on natural raw materials, which can even be broken down by the human body. Base materials include **indigo, DNA, beta-carotene, caffeine,** natural colors, and **glucose.** Electrically conductive circuits made of organic materials are printed onto a bioplastic film, enabling the creation of a sensor.

In another project, scientists at the University of Illinois manufactured biodegradable electrical components such as **solar cells, transistors,** diodes, antennae, and even simple digital cameras using thin layers of silicon, magnesium, **biocompatible silicon, and silk.** They were able to influence the decomposition time through the layer structure.

APPLICATION

Scientists predict that biodegradable sensors could be used in the food industry, for example for testing the freshness of bread, the ripeness of fruit, or the interruption of cold chains. **Natural electronics** would be ideally suited to use in toys too. Biodegradability could also enable the implantation of electronic components in the human body, in order to monitor the functions of the organism and metabolic processes and to document the progress of diseases.

Biodegradable electronics
(Source: Beckman Institute, University of Illinois and Tufts)

03

Recent years have seen an increase in the recovery of raw materials from waste products. For example, at present 90% of the roughly 200 million tons of extractive waste from the German construction industry can be kept within the economic cycle. In Central Europe, the recycling rates for paper and glass are over 80%, and even for plastics the rate of recycling stands at more than 60%. It is not only optimized waste management technologies that are leading to this particular upward trend, but also rising crude oil prices. Separating out high-grade plastics is becoming an important economic factor. Numerous commodity plastics command prices of more than €1,000 per ton on the global market. However, reports on plastic particles in the sea show that there is a great need for improvement in the recycling chains.

Designers and architects have repeatedly demonstrated in innumerable concept approaches and product designs that alongside the industrial utilization of raw material resources, used materials and waste products can also be employed as they are. Furniture made of old clothing, tableware consisting of used textiles, vehicle interiors composed of pressed newspapers, and wall mirrors constructed of old scaffolding planks are just a few outstanding examples of a development that highlights the increasing value of materials in society. Reusing materials not only saves resources, but also significantly lowers the energy required for material production. Moreover, by recycling high-grade metals and rare earth elements, industrial nations are reducing their dependency on Asian, South American, and African countries.

91

RECYCLING MATERIALS

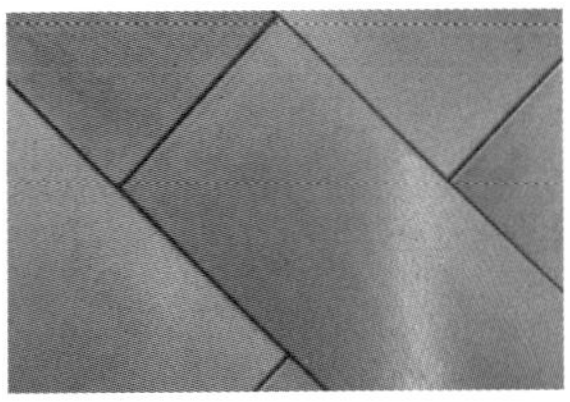

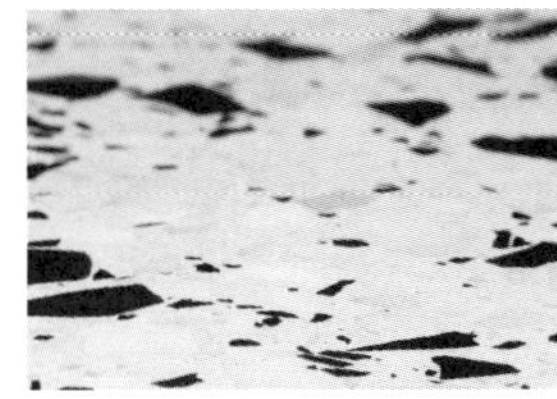

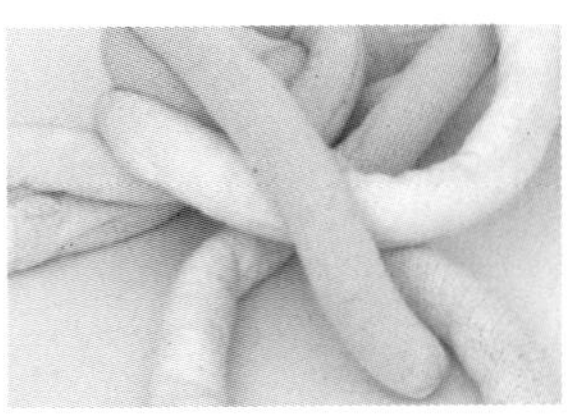

Rising prices for metal materials mean recycling processes are becoming increasingly important. This applies above all to particularly rare metals or to alloys, which have been hard to recover thus far.

Sustainability aspects recycling of hard-to-extract resources // energy saving compared with new production

SCRAP METAL MATERIALS

PRODUCTS

Bio-Luminum™

Distributed across the world there are thousands of "airplane graveyards." Due to the high volume of energy required to extract **primary aluminum,** recycling of this lightweight metal contained in the fuselage of planes is of great interest. With Bio-Luminum™, CoveringsETC has introduced to the market panels for floors, walls, and counters that come under particular stress, which are made from 100% **recycled aluminum** from aircraft. Until now it has been extremely difficult to recycle the aluminum alloys used in aviation. The US company has developed a recycling process for this very purpose, in which the scrap aluminum is converted into blocks which are then cut into slices. The resulting sheet material has a structure and look that appeal to designers and interior architects.

Recycling rare earths

The use of valuable metals from the range of rare earths is becoming ever more important in high-tech developments. These comprise 17 so-called "rare earth metals" such as **europium, dysprosium** or **ytterbium,** which are essential for the production of smartphones, wind turbines, and **fuel cells.** Due to China's domination of the world market for **rare earths** and the price increase in the recent past, recycling systems are currently being installed for the recovery of the metals from waste products. For the recovery of **samarium** and **neodymium** from **magnetic materials,** a process combining pyrometallurgical and wet chemical steps was developed at the Fraunhofer IFAM in Dresden. This made it possible to extract entirely pure raw materials for the production of new magnetic materials. There was great emphasis on reducing the need for chemicals in the process development.

Application of Bio-Luminum™
(Source: CoveringsETC)

Mining of rare earth elements (Source: US Geological Survey 2010, Saperatec)

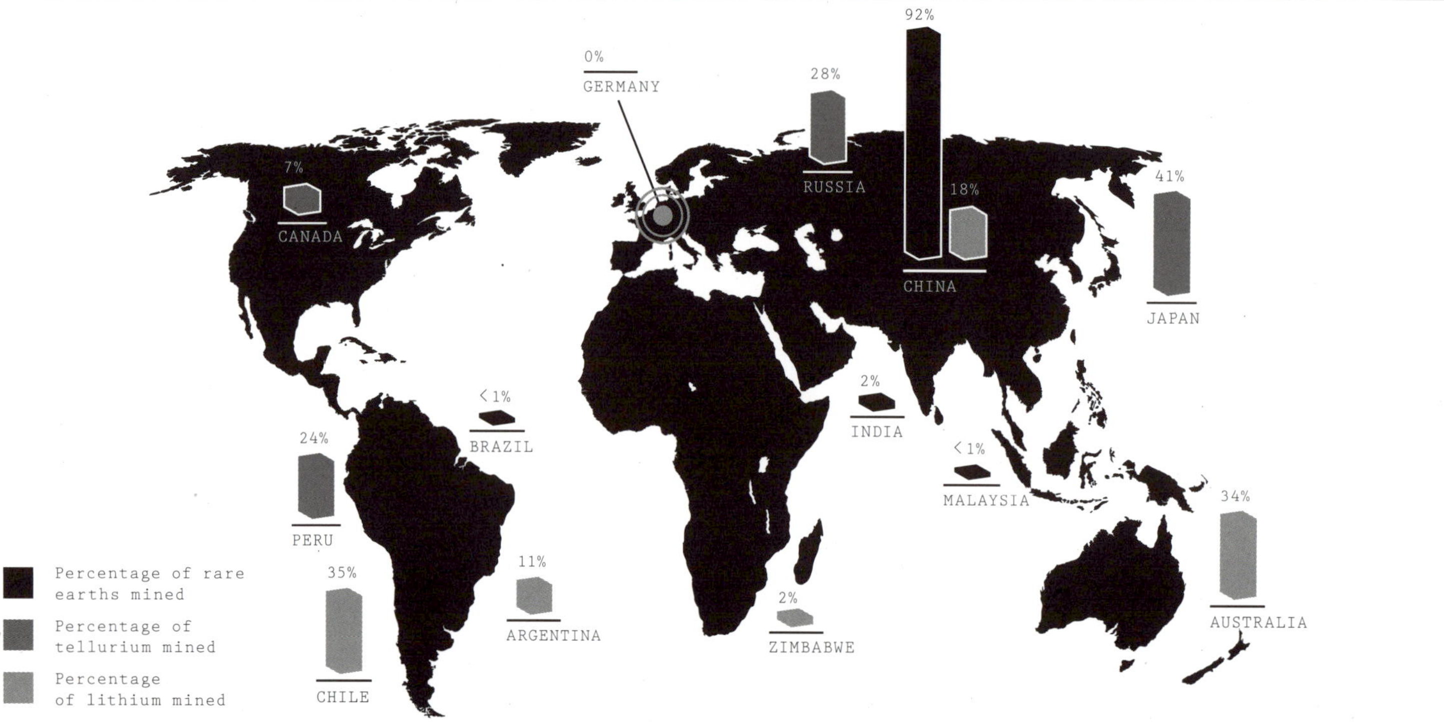

Sustainability aspects recycling of synthetic waste // upcycling concepts // energy-efficient production

WASTE PLASTIC MATERIALS

Thermoplastic synthetics are generally recyclable and can also be reused in a similar application or as a component of a recycled material. For this, the polymer waste must be collected and separated according to type. However the fact that, despite varied efforts, this still does not function to a satisfactory extent is borne out by the horrifying reports of the plastic waste that has built up over the years in the sea.

In 1997, scientists highlighted the problem of synthetic particles in the swirls of large ocean drift currents. The "Great Pacific Garbage Patch," a collection of an estimated 3 million tons of synthetic waste spreading over an area the size of Central Europe, in the north Pacific, has becoming increasingly worrying over the last few years, not least because the minute particles are consumed by living creatures and thus make their way into the human food chain. In some species of sea bird, turtle, and fish, the consumption of indigestible synthetic waste has already been proven to be a primary cause of death.

Wall made of POLLI-Bricks (Source: Miniwiz)

Packing from synthetic waste from the sea (Source: Method)

Since the synthetic particles remain suspended at depths of between 10 and 30 meters, and are distributed over large areas, disposal of the waste seems a virtually impossible task. Some governments have already banned plastic bags made of petrochemical polymers in reaction to the growing problem of waste.

A large quantity of the synthetic waste ends up in the sea via beaches. It was for this reason that an initiative entitled "Plastic Oceans" was launched with the aim of collecting synthetic waste on beaches and using it for recycling.

Ocean drift currents lead to the accumulation of synthetic particles

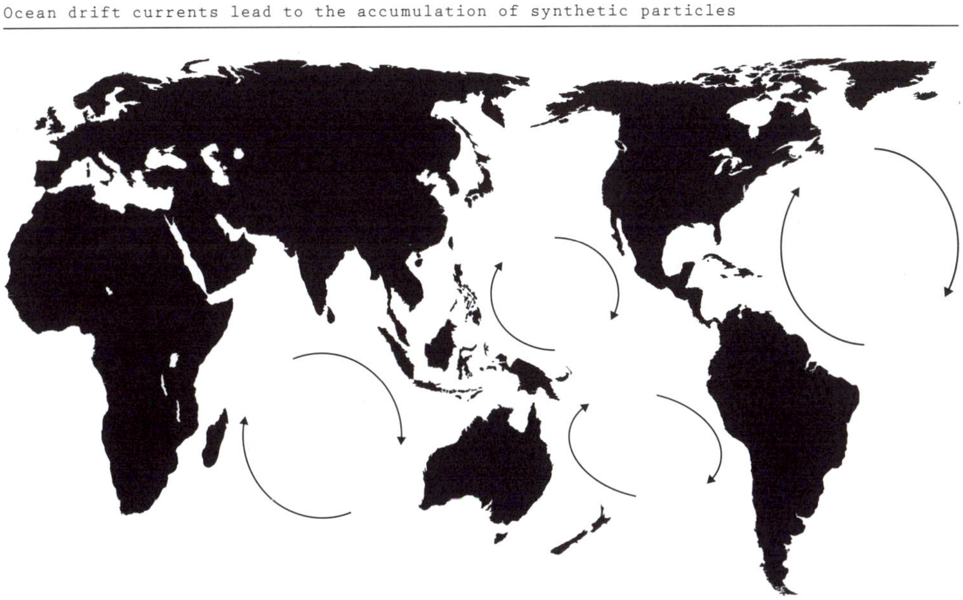

PRODUCTS

Ocean plastic bottles

In the fall of 2012, Californian company Method launched the very first product – a soap dispenser – to be made with waste plastic recovered from the sea. Method's founders have set themselves the task of demonstrating how environmental protection and economic success can be mutually dependent in a positive way.

POLLI-Brick™

Miniwiz from Taiwan has developed an architectural construction material made from **PET bottles.** These modular and particularly lightweight elements can be used to build structures for exhibition stands, dividing walls, fences, and ceilings. Since POLLI-Bricks are translucent, they make it easy to build back-lit structures. POLLI-Bricks also boast excellent noise and heat-insulation qualities.

Elastomer powder modified thermoplastics (EPMT)

The total global annual output of elastomer plastics (rubber) is around 22 million tons. Most of this goes into the production of tires. Because elastomers do not melt when subjected to heat, **waste rubber products** can be recycled at the end of their life cycle, either for energy recovery or in the form of small particles for surfaces including race tracks. An alternative technology for high-quality recycling of rubber waste comes from Fraunhofer UMSICHT in Oberhausen. Here, EPMT receptors were developed which contain up to 80% rubber waste and only 20% polypropylene. In a patented process, the waste rubber is first chopped into small pieces, then cooled with liquid nitrogen and ground into elastomer powder. This is then mixed with thermoplastic and additives in a melt-mix process. The result is a high-quality plastic that can be processed by injection molding and extrusion.

POLLI-Bricks recycling concept
(Source: Miniwiz)

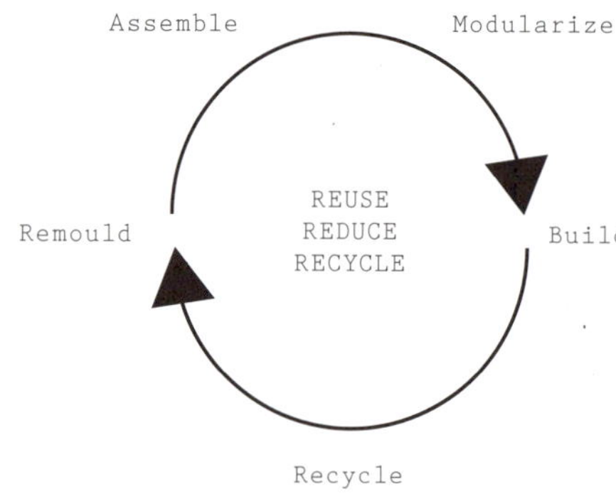

Every German citizen buys an average of 70 new pieces of clothing per year and disposes of used clothing in similar quantities. This produces 750,000 tons of used textiles every year, which can be recycled in various qualities. Alongside the classic recycling through used clothing collections and the reuse of fabrics from waste textiles in production systems, designers have developed some interesting approaches and business ideas over the last few years based on the recycling of waste textiles.

Sustainability aspects reuse of old textiles // increase in material and energy efficiency

WASTE TEXTILE MATERIALS

"Bis es mir vom Leibe fällt"
(Source: Elisabeth Prantner)

PRODUCTS

Stadtfund

The Berlin-based upcycling designers at Stadtfund have reacted to the problem of lost gloves with a remarkable business idea. They collect them, bring two of the same size together again as a pair, and sell them over the Internet. This way, they encourage people to mix and match and also raise questions about the mechanisms of the throwaway society.

Bis es mir vom Leibe fällt

Elisabeth Prantner has developed an extraordinarily simple form of textile recycling with her **"Veränderungsatelier" or "Change Studio"** in Berlin, and thus made it onto the winners' list of

Non-matching pairs of gloves
(Source: Stadtfund)

the national **Ecodesign Prize** in 2012. The collection, called "Bis es mir vom Leibe fällt" ("Until it falls off me") demonstrates ways in which constant repairing, redesigning, and customizing can be a principle for fashion design. The result is fashion that produces a particular dialogue between the textile and its wearer.

Furniture from old clothes

Tobias Juretzek was one of the first to come up with the idea of using old textiles for furniture building. As a result, his "Rememberme" chair consists only of old clothes like trousers, T-shirts, shirts, and blouses, which are pressed into shape under high pressure with the addition of a **resin mix.** After the hardening process, the result is a piece of furniture that preserves the individual qualities of the pieces of clothing, such as holes or lingering red wine stains, as a memory.

Textile designers Moa Hallgren and Lisa Spengler also produced a collection of **textile furniture objects** for their final piece at the Weißensee Kunsthochschule in Berlin, which consisted of fabric remnants, old clothes, and furniture that had been discarded as **bulk waste.** The objects, which were given the name **REMÖTIL,** were produced by hand, forging new links between furniture and textiles.

Textile bowls

The designer Kathrin Morawietz has made use of traditional turning techniques to produce a series of five bowls using a block of clothing held together with wood glue. These are surprising to the touch due to their textile-like feel and are particularly lightweight. Colored layers create lines in the pieces like veins in rock.

"Rememberme" chair, made from old clothes
(Design: Tobias Juretzek)

VEIO textile bowls
(Design: Kathrin Morawietz)

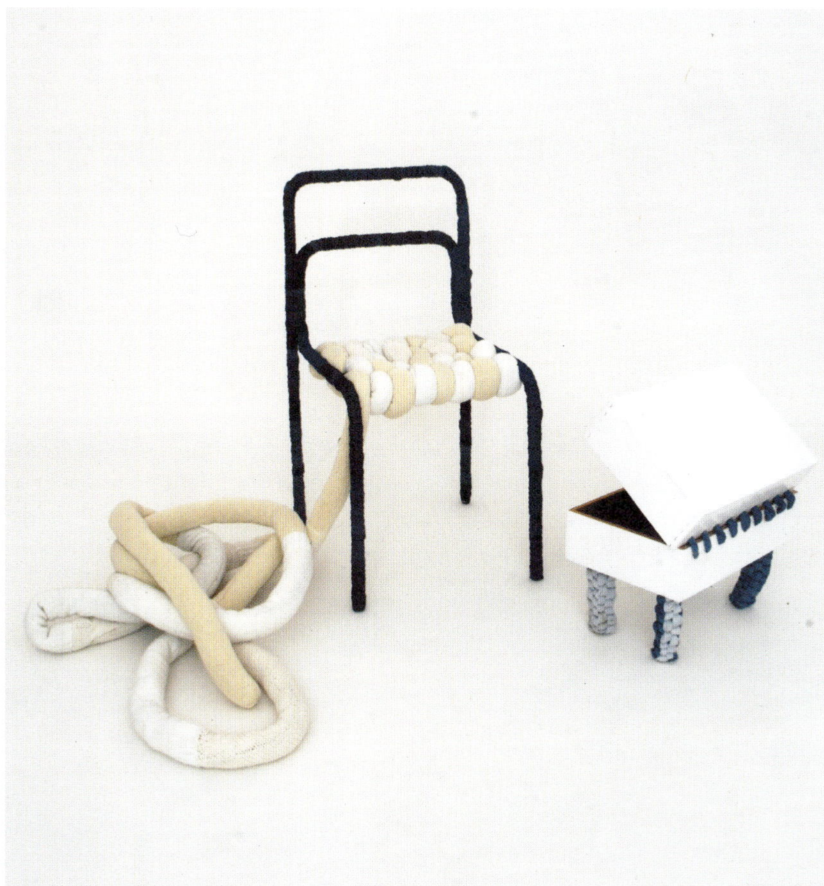

REMÖTIL – textile furniture objects
(Design: Moa Hallgren and Lisa Spengler)

Sustainability aspects use of wastepaper // material-efficient production and weight-saving

WASTEPAPER MATERIALS

Breaking down wastepaper and using the fiber pulp that is created, with the addition of water as a sculpting mass, is becoming ever more popular among designers. It is predominantly the ease with which it can be accessed and processed that makes it easy for creative types to put their design ideas into practice. The growing importance of waste materials for design and product development has driven the development of paper pulp and its processing.

PAPER PULP MATERIALS

DuraPulp

In the hunt for alternatives to oil-based synthetic materials, Swedish company **Södra** is exploring a combination of cellulose fibers and biopolymer polylactic acid, or PLA. The material consists entirely of renewable raw materials, is pleasing to the eye, and is sufficiently stable for furniture production. Preliminary seating modules and lampshades have already been produced from the material. The possibility of processing in an injection-molding procedure is being researched at **Fraunhofer UMSICHT** up to the end of 2013.

PaperLite®

PaperLite® is a thermoformable packaging material. It consists of wastepaper fibers and is biodegradable. It contains no **isocyanate** or solvents and can be used for food packaging. PaperLite® is lighter than conventional plastic packaging, easy to process and can be layered in **flexo printing.**

HI-MACS Eco Pulp®

This paper pulp is a surface material made of wastepaper fibers for the mineral material HI-MACS. By choosing the correct paper color, the manufacturer is able to create natural patterns without the use of pigments.

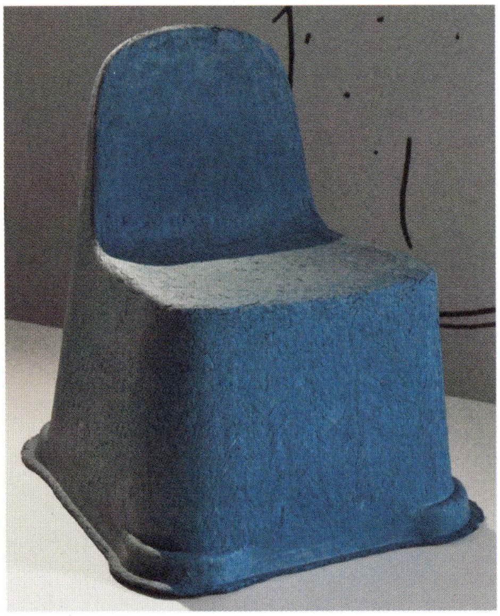

Seating modules made of DuraPulp
(Source: Södra Pulp Labs)

Surface material Eco Pulp®
(Source: HI-MACS)

PaperForms tiles

In Mexico and the USA, waste paper is used to produce tiles that are extremely lightweight and can be applied to the wall using double-sided sticky tape. The three-dimensional tiles completely transform a room. They are available in various patterns, textures, and colors, and are sold in packets of 12 with a format of 30.5 × 30.5 centimeters.

Pulp-based computing

At the Massachusetts Institute of Technology (MIT), an experiment was carried out to integrate electronic components, sensors, and actuators in the paper pulp. As a result, it was possible to produce so-called **"Electronic Paper Sandwiches"** and to create **paper loudspeakers, luminescent paper surfaces,** or electronic functions in packaging.

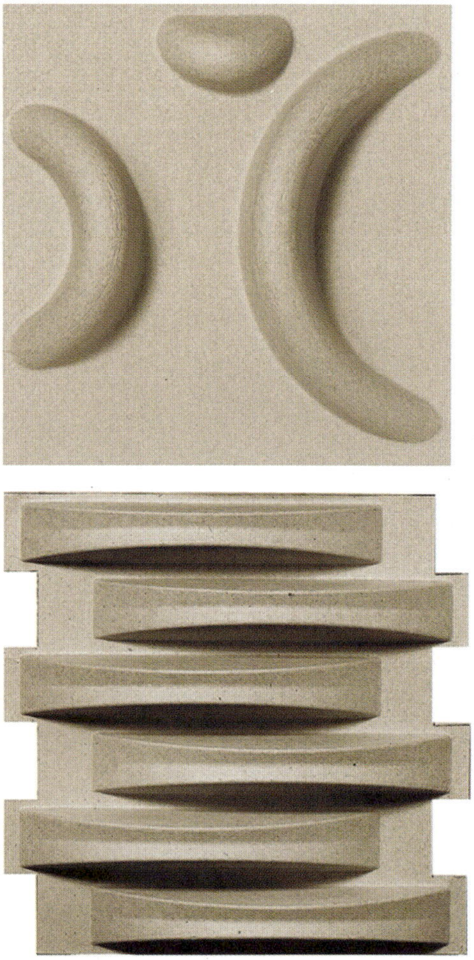

PaperForms tiles
(Design: Jaime Salm)

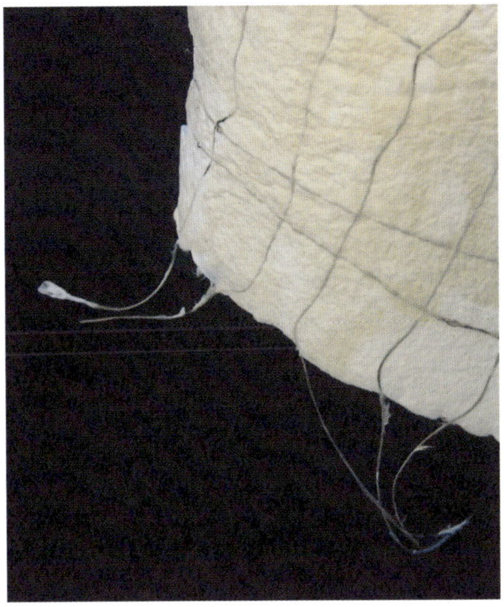

Pulp-based computing: Luminescent paper object Paperpulp (Source: MIT Media Lab)

NewspaperWood

A very interesting route has been taken by Dutch designer Mieke Meijer from studio Vij5 in the development of NewspaperWood. This involves the use of paper from old newspapers pressed together with a waterless glue so that, when cut open, it has lines resembling the rings in a tree trunk. The material is now being used as an **alternative to wood** in furniture items and home accessories. It can be sanded and milled, but can also be cut by laser. What is particularly striking is its low weight in comparison with natural wood.

Cartis paper furniture (Source: Cecil Karges)

NewspaperWood (Design: Mieke Meijer)

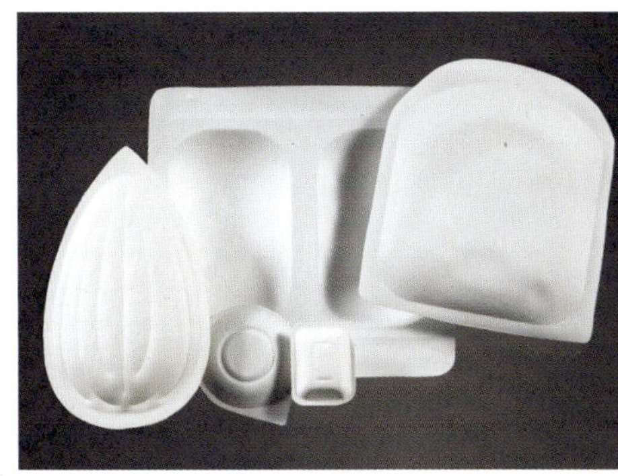

PaperLite packaging (Source: Flextrus)

Furniture made of NewspaperWood (Design: Mieke Meijer)

NewspaperWood in the interior of the concept vehicle ONYX (Source: Mieke Meijer)

Sustainability aspects recycling of long-life materials // long-term storage of carbon dioxide // reduced need for processing

WASTE WOOD MATERIALS

Over the last few years, there has been a perceptible trend in Central European interior and furniture design towards the use of materials with a natural character and imperfect surfaces. Obviously smooth and high-gloss surfaces have become dated and no longer reflect the zeitgeist. So brushed and sawn surfaces are currently very popular among architects and interior designers, as are objects and panels made of recycled wood.

One example is the designer Bauholz design from Münster, who specializes in creating furniture and fittings made of old scaffolding planks. The individual furniture items are produced by a manufacturing company and boast a unique style. The material comes exclusively from systems-oriented planks from German scaffolding, and the resulting room dividers, shelves, seating, and tables are manufactured according to customers' requirements.

PRODUCTS

Old Oak

Oak trees are particularly enduring and resilient and can live for over 1,000 years. In Central Europe in particular, **oak** – with its distinct shades – is extremely popular. Since 2004, Old Oak in Rosenheim has been using centuries-old and recycled oak to make parquet floors, floorboards, and design objects. The rough outer edges of the wood are used by designers to decorate walls, bars, or furniture surfaces. On request, flaws, grooves, and holes can have a silver, gold, or black infill.

Wonderwall

Wonderwall is the name adopted by a Dutch manufacturer of individually produced wall panels, which are assembled in a collage using a variety of old pieces of wood arranged in various depths. Each construction is realized according to the customer's particular needs and might consist of elongated hard wood planks in a dark color, or perhaps a variety of multicolored small wooden pieces combined. All the wall panels have one thing in common though: the natural, used character of the wood pieces used.

Collage made of old pieces of wood (Source: Wonderwall)

Wall made of wood from whiskey barrels (Source: McKay Flooring)

Wine cask parquet (Source: Scheba)

Old wood-veneer boards

Since 2012, veneer manufacturer Kaindl has included **wood-veneer boards** in its range, the veneers of which come from old wooden beams. This makes for greater naturalness and considerable ecological sustainability.

Wine cask parquet

Wine casks are produced from **oak.** Near Ingolstadt in Bavaria, there is a company that uses the wood from old **red-wine casks** to produce parquet floors or wall paneling. The wood has a fine-pored surface and has assumed the color of the wine after years of storage. **Wood from wine casks** forms only the surface layer of the parquet, however. It is glued onto an 8 mm middle layer of plywood, which has a 2.5 mm counter-layer of hardwood underneath.

Whiskey barrel flooring

At McKay Flooring in Glasgow, flooring is made using the **oak** from old **whiskey barrels.** The manufacturer guarantees that 3% of the boards even display the brand mark of the cask.

Table made of old scaffolding planks (Source: Bauholz Design)

Wall cladding using recycled wood (Source: Old Oak)

Sustainability aspects based on recycled material // reduction in energy consumption

MATERIALS MADE FROM RECYCLED CERAMICS AND GLASS

Since glass and ceramics are produced using a powder-based material, they are ideally suited to the recycling of leftover materials and waste. These are sorted, ground, and milled and can be returned to a chosen form in a thermal process. Alongside the recycling of material waste, manufacturers are looking increasingly towards reducing the energy needed for production. In modern production plants, the waste heat contained in the exhaust gases is returned to the production process and used for heating purposes. Reduction of water consumption is similarly significant in production.

ECOTECH®

In the production of **ECOTECH® porcelain stoneware,** the producer Floor Gres makes use of residual materials from **ceramic production** for 50% of the total volume. Along with its high proportion of recycled material, the structure and granulation are particularly striking.

Cristallino

Italian firm Trend Group is a producer of high-quality stone materials that are largely made from granulated recycled **glass.** One example is Cristallino, with a maximum format of 3,000 × 1,200 mm and a thickness of barely 7 mm for high-quality counters and work surfaces, which contain up to 70% **recycled glass**. The material also consists of around 10% **polyester resin**, into which colored pigments are mixed.

Eco-Gres®

Production of **porcelain stoneware** requires substantial quantities of water. If the water used in the production process can be reused, then the water requirement is reduced by up to 60%. CoveringsETC, the manufacturer of Eco-Gres®, has committed not only to the reduction of water consumption, but also to the recycling of waste material. In addition, the heat in the waste gas is recovered, which leads to a reduction in **overall emissions.**

EcoX

This concrete material comes from the USA and consists of 75% recycled glass. The extraordinary appearance of the material, which is available in sheets, lends itself particularly to use in furniture, dividing walls, shop fittings, and even sculptural objects. It is available in six standard colors.

Bottle Alley Glass

This British manufacturer specializes in the production of panel materials for furniture construction and interior design using **old glass bottles.** Bottles are sorted according to color, ground up, and transformed into a sheet material in the oven. After cooling, the recycled material can be cut to the required size, resulting in **tiles** with edges of a maximum of 300 mm, countertops 10 mm thick or splash guards of up to 1100 × 750 mm. **Light diffusers** made of **recycled glass** are also supplied.

GAIL Architektur Keramik

Nowadays some ceramics producers guarantee they will reduce resource consumption by reusing a proportion of the waste that results from the production process in manufacturing. GAIL Architektur Keramik duly talks of reusing at least 5 to 20% of the ceramic waste and the active recycling of 100% of residual material from across the ceramic production process. In addition to this, excess heat energy from the waste gases is reincorporated into the process or is used to heat the office building.

EcoX consists of 75% waste glass (Source: Meld USA)

Countertop made from old glass bottles (Source: Bottle Alley Glass)

Sustainability aspects based on recycled materials // reduction of carbon dioxide emissions // recovery of valuable resources

CONSTRUCTION MATERIALS MADE FROM WASTE

Taken together, cement factories worldwide emit in excess of a total of one billion tons of carbon dioxide, which equates to 5% of global CO_2 emissions. With the aim of reducing greenhouse gas emissions, numerous new construction materials made using recycled materials have been developed over the last few years. Scientists are also trying to lower the temperature needed for cement production in order to reduce CO_2 emissions.

PRODUCTS

Celitement®

One such construction material is Celitement®, which resembles **Portland cement** in its characteristics and can be produced at less than 300 °C (normal temperature is 1,450 °C). It is based on hydraulically active **calcium hydrosilicate** and takes around 50% less energy to produce. The procedure was realized at the Karlsruhe Institute of Technology (KIT) due to research into cement on a nanometer scale with the use of **synchrotron radiation.**

Syndecrete

Syndecrete is a cement material developed by the architect David Hertz with the use of recycled materials. It consists of bulking agents such as metal cuttings, pieces of glass, plastic granulate, or old wood and is supplied in small formats. Compared with conventional concrete, Syndecrete weighs half as much but has twice the compressive strength. Its Italian **terrazzo**-style appearance is particularly striking.

Eco-Cem

This concrete material is suitable for use as paneling for floors, work surfaces, and walls and consists of up to 80% cement and 20% recycled **cellulose fiber.** Its moisture and gas permeability is supposed to have a positive influence on the internal environment. Eco-Cem is available in eight different colors.

Eco-Terr

The Eco-Terr boards and tiles produced by American producer Coverings Etc consist to a great extent of residual stone from granite, marble, or river gravel. In addition, glass particles are added to the material, which give it a terrazzo look on floors, surfaces, or wall paneling.

Eco shingles
(Source: Seneca Shake)

Eco-Cem in use
(Source: Eco-Cem)

Eco-Terr panels
(Source: CoveringsETC)

Biodämm

This **hollow block** material consists of almost 45% foam glass, around 5 – 15% residual substances resulting from concrete production, plus a binding substance and gravel as a further filler. It offers sufficient noise and thermal insulation properties for the construction of internal walls. Energy consumption in production is significantly reduced thanks to an oven technique with **heat recovery.**

Eco-shake

Eco-shake **shingles** are made entirely from recycled plastic with cellulose fiber reinforcement and were developed as an alternative to wooden shingles. They offer not only excellent weather resistance but are also particularly lightweight. For **roof cladding,** the boards are available in various colors and formats. Given to the product's longevity, the manufacturer is able to offer a guarantee of 50 years.

Rice cement

Taiwanese company Miniwiz has developed a procedure to extract silicates for cement production from agricultural waste such as **rice husks.** The silicate is amorphous and, with a proportion of 98%, contains significantly more silicon dioxide than silicate dust or fly ash. The new technology means less cement is needed in concrete production, which significantly reduces carbon dioxide emissions.

Urban restructuring and mining robot

In 2012, industrial designer Jan Meissner caused quite a stir with his future scenario for the recovery of construction materials in large cities and megacities. He devised a semiautomatic system for dismantling high-rise buildings using cutting robots, flame cutters, and shredders, which dismantle the stories in a systematic fashion, returning the waste materials to the **materials cycle.** A pipe system on the external walls of the building transports the waste material to the ground, where it is shredded and sorted into glass, concrete, steel, and residual waste. The X-frames and the protective cases hung on them gradually get closer to the ground as demolition progresses.

Urban Mining / Reconstruction robot system
(Design: Jan Meissner)

LIGHTWEIGHT CONSTRUCTION MATERIALS

04

Today, ever scarcer resources are compelling manufacturers of materials to enhance the efficiency of those they offer and to support sustainable product and industrial design by lowering the materials' weight – this can take the form of folds, crystal-like structures, sinus-shaped honeycombs, ring elements, or foamed sections. This applies to the construction industry just as much as it does to the aerospace and automotive industry, where the success of electromobility will crucially hinge on reducing vehicles' weight. Lightweight materials not only cut the amount of energy required for transport, but are also simpler to assemble and drastically reduce production and assembly outlays. It thus comes as no surprise that in the years 2002 to 2007, demand for lightweight construction materials soared by 300 percent.

This trend will become more pronounced given the world's burgeoning population and economic developments in the emerging markets. Given rising prices for products derived from crude oil and the consequent search for regenerative materials that can function as substitute bases for plastic, wood is evidently becoming increasingly important, and by extension lignocellulose. As a result, the wood-processing industry has developed innumerable types of lightweight construction panels that do justice to the need to spare resources and deliver lower material inputs. We are also seeing innovations in the field of fiber-reinforced materials. And, in summer 2012, German scientists presented a new carbon-based world record holder in terms of weight: aerographite.

LIGHTWEIGHT CONSTRUCTION MATERIALS

Due to the greater use of lightweight construction materials such as aluminum or fiber-reinforced plastics, classic construction and structured steels face tough competition. Manufacturers such as ThyssenKrupp and scientists at the Technische Universität Bergakademie Freiberg are working vigorously to reduce the weight of steel and are now offering lightweight steels.

Properties alloying with light metals // enables thin walls // high strength // excellent toughness

Sustainability aspects increases material efficiency // energy saving in the automobile industry

LIGHTWEIGHT STEEL

MATERIAL CONCEPT AND PROPERTIES

Lightweight steels can be produced, for instance, by means of adding metals with a low density (including **aluminum**). In steels with a high concentration of **manganese** (FeMnAlSiC), a weight reduction of 15% is possible. Simultaneously, alloying with other metals permits a significant enhancement of the mechanical qualities, to realize structures with a thin wall. At the Institute of Iron and Steel Technology in Freiberg, for example, researchers developed lightweight steels with increased strength and simultaneously excellent toughness.

Thin-wall cast steel (Source: EDAG)

APPLICATION AND PROCESSING

Lightweight steels are primarily interesting for the automobile industry since lower weight also cuts fuel consumption. The rule of thumb is that in normal usage a weight saving of 100 kg can achieve fuel savings of between 0.3 and 0.4 liters. In developing the "Future Steel Vehicle" at EDAG in Fulda, engineers employed innovative lightweight steels with strengths up to 2,000 megapascal for the bodywork and in the process modified and updated production techniques such as press-hardening and roll-forming. In addition, components were realized in cast steel and structures hardened selectively. Due to lightweight construction, weight savings of 188 kg can be achieved.

In recent years, new concepts and manufacturing processes have been developed to reduce vehicle weight and to replace relatively heavy metal components with lightweight composite materials made of plastics.

Properties thin-walled panels // fiber-reinforced plastics // low weight // high strength // simple processing when heat is applied in the thermoplastic matrix // integration of connection elements and bearings

Sustainability aspects energy saving in transport and operation // material-efficient production

ORGANIC SHEETS

MATERIAL CONCEPT AND PROPERTIES

Organic sheets are thin-walled panels and structures of **fiber-reinforced plastics,** which demonstrate similar mechanical strengths to metals. The matrix system is described as organic because the binders are produced using petrochemistry. Customary fiber materials are **glass, aramid** fiber or carbon fiber. For high-performance composites, continuous fibers or weaves are almost exclusively chosen. Linen, twill weave or **atlas weave** are primarily employed as weave types. Depending on the desired quality, organic sheets can be realized with a **thermoplastic** (e.g., TPU, PP, PA) or thermosetting **duroplast matrix.** It is also possible to apply organic sheet as the top layer to a high-strength foam core. This permits a weight reduction of up to 60% in two-dimensional vehicle components. As they are composed of different materials, organic sheets are not easily recycled.

APPLICATION AND PROCESSING

The great advantage of using a **thermoplastic matrix** is the malleability under heat. It means components can be easily bent or pressed into shape to realize diverse applications in the fields of sport, orthopedics, or vehicle construction. Organic sheets are of special interest to the automobile industry as they can be processed using the equipment already available. It is also possible to integrate follow-up systems, connection elements, or warehouses into the matrix system. Organic sheets can be bonded, welded, and varnished. **Aramid fibers** are employed chiefly for impact-protection solutions.

Roof of organic sheet for the Smart ForVision (Source: BASF)

Organic sheet for use in the automotive sector (Source: BOND Laminates)

Types of weave used in the production of organic sheets

Properties lower material consumption // adjustable fiber density // low weight // closed-pore cavity structure // as far as possible purely organic // lightweight middle layer // low material consumption

Sustainability aspects helps reduce toxic fumes // lowers energy requirements

WEIGHT-OPTIMIZED TIMBER MATERIALS AND REPLACEMENT MATERIALS

Although Germany is one of the most densely wooded regions in Europe, there is expected to be a bottleneck in procuring wood in future. To protect existing forest land and create timber materials with constant properties in the 1990s, a wide variety of material panels were developed that are based on waste from the timber industry and use synthetic adhesives made of urea-formaldehyde or polyurethane (PUR) resins. Twenty years after their development, the increasing importance attached to sustainability has resulted in numerous advances in innovations to further reduce the amount of material and to counter the toxic impact of synthetic adhesives.

PRODUCTS

Dascanova

A remarkable technology to reduce the percentage of wood in plywood panels comes from Dascanova in Austria. The start-up firm is able to specifically regulate the concentration of fibers in a panel to modify the density and strength precisely to a particular application. This makes it possible to make plywood panels using 30% less material. This reduction has a follow-on effect, also cutting down the amount of resin needed and the energy required for drying.

BalanceBoard

A new board material has been launched that is some 30% lighter than conventional plywood. This is achieved by adding corn or wheat starches, which foam when heated and form cavities in the wooden mass. The **BalanceBoard** can be processed using conventional techniques and can be recycled. Despite being less solid, tests have shown that boards with a middle layer of hollow space demonstrate higher strength.

Rye fiberboard

The method of foaming substances containing starch was also used in developing the rye fiberboard ROFA by the IGV Institute for Grain Processing. Foamed rye flour makes up 50% of the finished product, while 40% is wood fiber. Rye fiberboard lends itself particularly to landscaping and erosion protection as it can absorb high amounts of water, namely 15 liters per square meter.

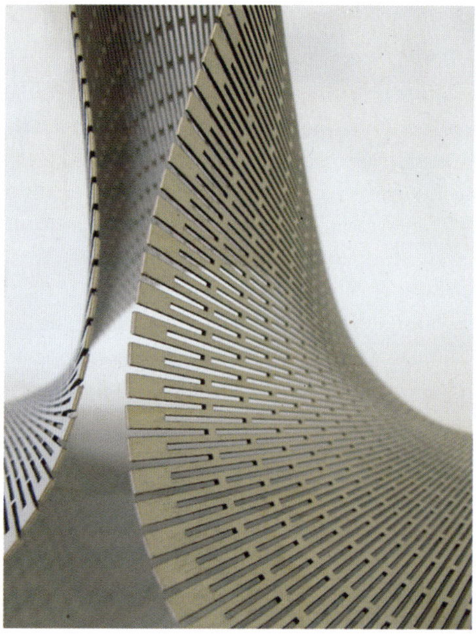

Incisions can reduce the weight of wood-based materials (Source: Dukta)

EcoSystem natural fiberboard for making furniture (Source: UdK Berlin, System 180)

Cavity structure based on foamed cornstarch

Plywood with a soya-based binder (80% soya, 20% synthetic cross-linker) (Source: Naporo)

EcoSystem

A further example is the natural **fiberboard** EcoSystem, which was developed at the Berlin University of the Arts (UdK) in cooperation with furniture producer System 180. Renewable fiber materials from agricultural waste were used in its manufacture, and rather than using urea-formaldehyde or PUR resins employed in customary wood materials, an organic plastic keeps the board in shape. The honeycomb-like support structure also reduces the amount of material needed, and consequently the weight.

Naporo organic plywood

A similar approach was pursued by Naporo GmbH in developing an organic plywood. This consists chiefly of natural fiber shavings held together using a soya-based adhesive. To ensure strength when exposed to moisture, a small amount of synthetic cross-linking agent **is added to the adhesive.** Both the adhesive and the entire plywood board are free of **formaldehydes** and assigned to the emission category E0.

Dendrolight®

Dendrolight aims to achieve comparable mechanical durability combined with markedly lower material consumption. The middle section is made of especially profiled spruce or pine boards that are bonded on each other at right angles. This makes it possible to reduce the weight by approximately 40%.

HeiLight

While many developments involve trying to reduce the amount of wood by using renewable raw materials, other manufacturers employ honeycomb structured central sections in a **sandwich technique.** In the newly developed HeiLight, wood veneer strips are deployed in place of the customary paper honeycombs. These strips can be arranged to suit the load requirement. Vertical veneer fibers, for instance, are more suited to applications where there is a high load requirement, whereas horizontal fibers provide favorable bending and rupture shear strengths.

Lisocore

This lightweight board can be made from renewable raw materials and is based on a double-curved shell structure, which makes the material up to 70% lighter while retaining virtually comparable static qualities. This makes it especially interesting for the areas of shipbuilding and the automobile industry, where it can help to cut energy consumption.

Dendrolight middle layer of profiled spruce and pine boards (Source: Dendrolight)

HeiLight central section of wood veneer strips (Source: Schotten & Hansen)

WEIGHT-OPTIMIZED STRUCTURED AND HONEYCOMB CONSTRUCTIONS

Properties low use of material // sandwich construction // hexagonally structured // sinusoidal // diamond-shaped // reinforced with coating layers // load-bearing 3D mesh structures // rigid core structure

Sustainability aspects resource-efficient material and structural solutions // simple assembly and processing

Working with honeycomb structures and structured elements produces material solutions with high stability and requiring less material. In recent years, there has been a marked increase in the use of such constructions. For some time, honeycomb carton structures have been deployed both in lightweight core elements in exhibition-booth manufacture and in shop-fitting, but also the automotive and aircraft fields. New solutions are now emerging that markedly improve material efficiency and reduce weight.

PRODUCTS

Sinusoidal honeycomb panels

SWAP makes **paper sandwich systems employing a sinusoidal honeycomb structure with excellent shape-retention qualities.** They consist almost entirely of waste paper and can be recycled after use. They stand out particularly for ease of handling due to their low weight and their excellent damping properties. In composite systems, cardboard honeycomb **structures** serve to integrate covering materials such as timber layers, metal and plastic panel boards, or acoustic felt.

HexFlex

The core of this new paper-based board consists of **Hexacomb,** a honeycomb carton **structure** with a hexagonal shape, and a covering layer of gray card or **Kraft paper** on the one side and a **cellulose fleece on the other.** It can be bent and reshaped without too much effort.

Nidacell®

The brand name Nidacell is applied to an extrusion process for making physical vapor deposition (PVD) **or PVD honeycomb** materials for **sandwich core structures.** Production allows for varying cell structures (hexagonal, rectangular, diamond-shaped, rice-grain shaped) and sandwich layers between 5 and 30 mm. Due to the thermoplastic properties, plastic covering layers are easily applied. Compared with solid materials offering similar stability, a 40% reduction in mass is possible.

Elybond®

The Elybond lightweight honeycomb structure is based on fiberglass-reinforced **polypropylene,** which makes it especially light and waterproof. One or several composite fiber layers are reinforced with external covering layers. The mechanical qualities can be tailored to suit the application by varying the number of layers and weave type. Standard is a three-layer polypropylene (PP) fiberglass-reinforced version with high tensile strength and pressure resistance. Expansion under heat is comparable with the performance of aluminum.

Octamold

A three-dimensional mesh structure capable of absorbing loads for core materials in sandwich structures, Octamold is based on a truncated rhombic dodecahedron, and represents an ideal relationship between surface and volume. As happens with the froth of soap suds, a balance of forces is established and the space completely filled. By connecting at least two levels, a layer of whole truncated octahedrons is produced, in which all rectangular surfaces represent joining and connecting surfaces.

3D-Core

3D-Core is an ideal core material for making lightweight structures in boat-building and aviation. This is due to the foam matrix of hexagonal elements, which are connected to one another by fine slats, adapting easily to slight curves and in doing so facilitating the absorption and deflection of thrusts, pressure, and bending forces. Production processes can be improved considerably vis-à-vis classic **laminates.** The number of layers and resin systems can also be reduced appreciably.

Luxpanel

Honeycomb panels of aluminum have proven to be especially valuable in industrial applications due to their high rigidity and bending strength. For example, the Luxpanel, with its **hexagonal** structured core (**cell width** 9.6 mm), offers special mechanical qualities combined with a low weight. Compared with customary solid materials of wood, steel, or concrete, a weight reduction of up to 60% is possible. Since a viscous adhesive is employed to produce the composite structure, Luxpanel is easily processed and also has high damage tolerances.

Tripan

These lightweight components are sandwich constructions whose core consists of a pressure-resistant and bendproof **aluminum honeycomb** or hard-foam core covered in layers of aluminum, steel, or plastics such as ABS, GFK, CFK or HPL. They are characterized by a low area weight combined with high durability. Hard-foam cores also have excellent heat insulation properties. The manufacturer can tailor the product to suit the application for which it is needed.

Luxpanel aluminum honeycomb panel (Source: Luxpanel, photo: Diana Drewes)

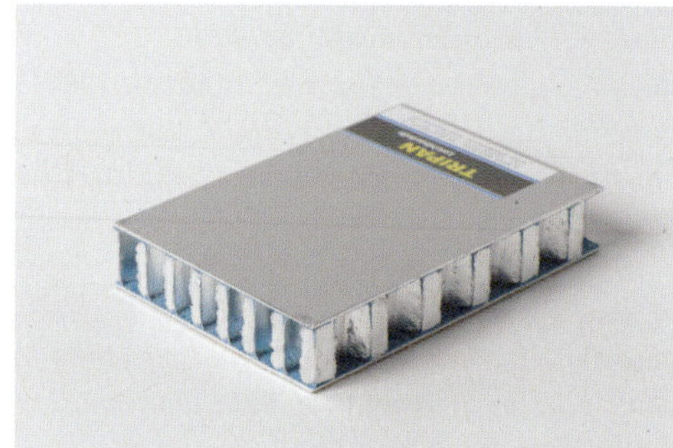

Aluminum honeycomb structures covered by a metal layer (Source: Tripan)

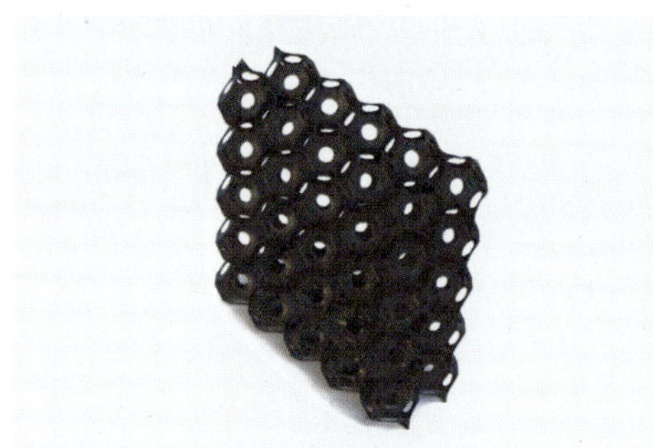

Octamold core material (Source: Octamold)

Weight-optimized sinusoidal honeycomb panel for exhibition stands (Source: SWAP)

WavCOR

Three triangular lightweight **structured cores by** American manufacturer ECOR consist entirely of **cellulose fibers.** This means they are completely recyclable and ideally suited for manufacturing furniture, exhibition booths and displays. There is an excellent relationship between rigidity and weight. The material can be easily coated and can be varnished or painted.

Borit

The Borit honeycomb panel is based on two honeycomb sheets with quadrangular, hexagonal, or octagonal elements, which can be fixed at a defined distance exactly above each other. Depending on the choice of material, different joint techniques can be employed, allowing solid joint components with large **contact surfaces to be realized.** The panels have an especially rigid core structure, are especially suited for applications in the construction and vehicle industries, but equally for exhibition architecture, furniture, or the packaging industry.

Dukta

Dukta also enables the flexible use of wood. By making incisions in the material it is rendered more flexible, permitting organic shapes for interior fittings and furniture design. The incision technique can be applied to timber materials such as medium-density fiberboard (MDF), oriented strand board (OSB), or plywood. In addition to enhancing its appearance, the perforations lend the material sound-absorbing properties.

Structural skin

Another exciting styling solution using sheet metal comes from Austria, where for the SKIN exhibition a collection of laser-cut panels were realized that are based on the geometry and symmetry of the circle and the triangulation of dots in the three-dimensionality of space. By pressing on the structures, the user can transform the structures from the two-dimensional to the three-dimensional.

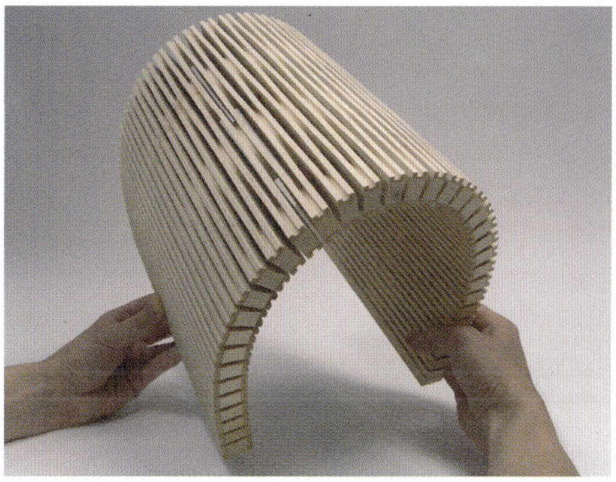

Flexible perforated timber (Source: Dukta)

Honeycomb sheets with hexagonal elements (Source: Borit)

Properties low use of material // comparatively high weight-bearing capacity // organic styling // wood/textile and wood/metal combinations

Sustainability aspects resource-efficient construction solutions // simple production

FOLDING LIGHTWEIGHT STRUCTURES

In next to no time, Takuo Toda folds a piece of paper into a working flying object. He is in the Guinness Book of Records for making a paper plane that remained airborne for 29.2 seconds: a world record. The chairman of the Origami Airplanes Association has crafted over 700 folding designs for paper airplanes, and impressively demonstrates the lightweight potential in folded structures. The Japanese man even plans one day to throw a paper plane out of the International Space Station and have it land safely on Earth. To ensure the project's success, the robust paper made of long sugar-cane fibers will be chemically treated and should take three days to sink slowly to Earth.

OTHER FOLDED STRUCTURES

Foldtex

This building material consists of at least two layers: a stiff plywood layer (1–6 mm, making up 60–80%) and a flexible fabric coating (15–30%). If Computer Numerical Control (CNC) milling cutters are used to cut into the base layer, the material can be transformed into a foldable construction held together by the textile. Depending on the choice of material for the flexible base layer, the folded construction can be imbued with additional properties.

Foldcore

Foldcore is a general term describing a technology for the stability-promoting **folding** of core structures for **sandwich materials** without the assistance of incisions or bonding or permanent deformations. It results in regular and irregular **patterns** with higher **load-bearing capacity** in differing materials for applications ranging from custom-made through to mass production. Foldcore is capable of folding paper in such a manner, for example, that up to a ton can be supported by just 10 g of paper.

PreBeam®

These flexible sections enable the efficient realization of organic design or curved surfaces. They consist of a folded **metal structure** of 0.15 mm thick **aluminum**. topped by a layer of plywood (4 mm), aluminum (1 mm), high pressure laminate (HPL) (1.2 mm) or cardboard (2 mm). Particularly when working with **birch plywood**, the use of screws, clips, or pins is possible. PreBeam sections are available in a maximum length of 2.65 m. Typical application areas include exhibition and trade fair design, stage-sets, and interior design.

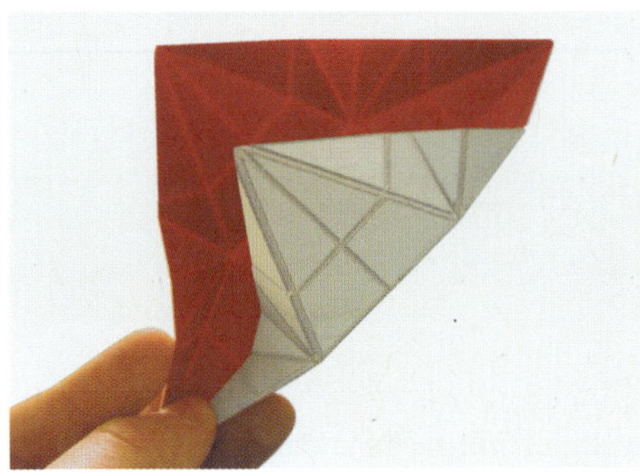

Foldtex made of plywood and textile coating (Source: Timm Herok)

Foldcore sandwich core material (Source: Foldcore)

Folded structure using Foldtex (Source: Timm Herok)

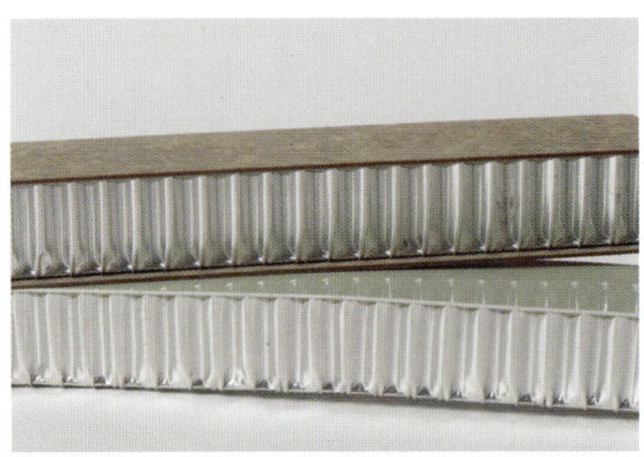

PreBeam section for organic mold-making

Properties possibility of material reinforcement // comparatively high load-bearing capacity // free-form structures possible // high damping

Sustainability aspects less material used

LAID SCRIM STRUCTURES

Working with laid scrims is one of the simplest ways of reinforcing materials with fibers. The individual fiber strands are arranged parallel in individual layers and fixed with binders or bonding agents at the crossing points. They are used in situations where woven materials are too large and where warp knits, due to their mesh structure, would be too flexible.

PRODUCTS

BafaTex laid scrims

BafaTex offers various applications for the construction, industrial, and packaging fields. Laid scrim structures are used, for example, as sandwich constructions for sports equipment, for reinforcing body armour, in **textile-laminated structural elements,** for furniture construction, or as inserts for vapor diffusion materials in the construction industry.

Bcomp Power Rib

The Swiss company Bcomp has developed a technology for the integration of twined fibers into a **flax fiber composite** with which greater rigidities can be achieved for applications in vehicle construction, the energy industry, and for outdoor items (20–30% higher than with glass-fiber reinforced plastic [GRP]). During production, the yarn is placed in a controlled way so that the mechanical properties can be set appropriately for the application. The patented technology in this process makes use of the increase in rigidity that the fibers are given through **twisting** during yarn production. In addition, each type of fiber and the thickness of the yarn influence the quality of the application.

splineTEX®

In Austria, a composite made of fiber-reinforced plastic with weatherproof qualities has been developed for the realization of load-bearing **free-form mesh structures.** These can then be used for the creation of free-form furniture, luminaires, and trade show stands, as well as for facade and construction elements, without the use of costly molding tools. After being shaped, splineTEX® structures are reinforced with resins or stabilized using connecting elements in the required geometry. In addition, the structural elements can be combined with other materials to create closed surface materials.

BafaTex filter non-woven material
(Source: BafaTex)

Power Rib fiber structure
(Source: Bcomp)

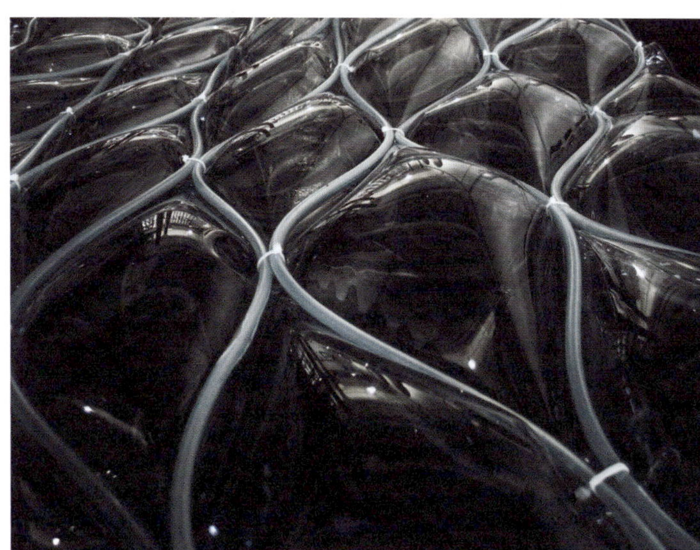

splineTEX® structure
(Source: splineTEX®)

INFRALIGHT CONCRETE

Properties dry density of less than 800 kg/m³ // usable as exposed concrete // low lambda value // addition of air-entraining agents // self-compacting

Sustainability aspects good heat insulation // efficient use of construction materials

Infralight concrete refers to a particularly dense concrete material which, with its good heat insulation properties, can be used as fair-faced concrete.

MATERIAL CONCEPT AND PROPERTIES

Infralight concretes have a dry density of less than 800 kg per cubic meter. **Air-entraining agents** and additives are mixed into the concrete, which improve the heat-technical properties of the material so that it can be used in **monolithic fair-faced concrete structures** with no further heat insulation required. One example is a development by Mike Schlaich at the Technical University of Berlin. Here, an **expanded clay** was added to an infralight concrete, so achieving a dry bulk density of 760 kg per cubic meter and a lambda value of just 0.181 watts per meter and Kelvin with an external wall thickness of 500 mm. To avoid shrinkage cracks and reduce the corrosion risk, the concrete material was reinforced with **fiberglass rods.** This way the formation of **heat bridges** could be avoided.

APPLICATION

Infralight concrete increases material efficiency in construction because single-sheet exposed concrete architecture becomes possible without the need for costly heat insulation. Its self-insulating properties make infralight concrete suitable for finished parts too. In 2012 a large-format test wall was awarded the Holcim Foundation's Innovation Prize.

Detached house made of infralight concrete with the addition of Liapor expanded clay in Berlin's Pankow district (Source: TU Berlin)

Finished wall construction made of infralight concrete (Source: TU Berlin)

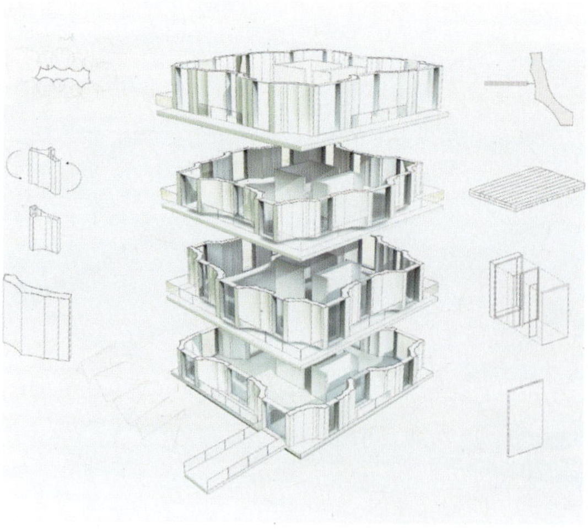

Smart Material House (Source: Barkow Leibinger Architects)

FIBRATED CONCRETE

Properties low wall thickness // delicate structures // low weight but good load-bearing capacity // organic design // simple to use // longevity // use of renewable raw materials preferred // exposed concrete look

Sustainability aspects efficient use of materials // long-lasting and weather-resistant // sparing use of resources

The use of glass, aramid, and carbon fibers as materials for reinforcement is long established, particularly in vehicle construction and the sports industry. The relatively new notion that carbon fibers are particularly suitable for realizing very delicate concrete structures for the building industry or furniture construction is now inspiring architects and manufacturers in equal measure to create new developments and designs. The use of natural fibers for the reinforcement of concrete is currently being researched.

PRODUCTS

Carbocrete®

In 2012, the SGL Group used an Open Innovation competition as an occasion for the first presentation of Carbocrete®, a new lightweight compound material made of carbon fibers and concrete, which is particularly suited to design and architecture applications. Carbocrete® is as sturdy as **reinforced concrete,** but up to 75% lighter and longer-lasting. Compared with construction elements made of reinforced concrete, the same parts made of Carbocrete® cannot rust, which makes them particularly suitable for contact with water. The material is easy to work, can be draped, and has a high **ductility.**

Concrete Canvas

With Concrete Canvas, a UK manufacturer has managed to combine the flexibility of a textile with the sturdiness of concrete. Originally designed for military purposes with the idea of erecting stone buildings in crisis-hit areas in the shortest possible time, the **spacer fabric** filled with **dry concrete** offers endless possibilities for designers. It can be flexibly molded and hardens completely on contact with water in the course of a day. Concrete Canvas is available as a rolled material in various thicknesses (5–13 mm) at affordable prices. In industrial applications it is being used for the reinforcement of hillsides, for example.

betoShell® textile-reinforced concrete

The betoShell facade system consists of facade plates only 30 mm thick, which are equipped with 3D **textile armoring** and can resist wind pressure of up to 1.80 kilonewtons per square meter. SITgrid® is a new **spacer textile** made of **glass fibers,** for the production of sheet materials using plastic or concrete, with an extremely high load-bearing capacity and a particularly low dead weight. A production line permits the manufacture of **concrete facade panels** in large dimensions of up to 6 × 3 m. The low weight is advantageous when it comes to transport and assembly. In comparison with steel-armored facade elements, the production of betoShell uses up to 80% less in resources. Concrete facades can be equipped with various colors of **aggregate grains** and can also have lined textures, smoothed or acidified looks.

Concrete slab with integrated carbon weave (Source: Paulsberg)

betoShell facade elements (Source: Hering)

Bamboo concrete

In Asia, in particular, **bamboo** has traditionally been used in construction. On the one hand, this is due to the plant's rapid growth. On the other, the material has particular qualities of sturdiness and performance characteristics that make its use ever more appealing for Western architecture too. Under the leadership of Dirk E. Hebel, a team at the Future Cities Laboratory in Singapore is currently looking into whether bamboo might be used instead of steel to reinforce concrete. Water absorption, shrinkage, durability, and the infiltration of the natural material by fungi are just some of the challenges the scientists are having to tackle.

Concrete wallpaper

Anyone who is not able to construct a building from scratch but still wants a concrete look can copy the aesthetic of **fair-faced concrete** due to concrete wallpaper. The wallpaper is a dimensionally stable **fiber woven** that can be easily applied to interiors. The material is also suitable for short-term use externally.

Exposed-concrete look using concrete wallpaper (Source: betontapete Berlin)

CNT-REINFORCED MATERIALS

Diamond, graphite, soot: all these materials are made of carbon. Their properties have been known for centuries. Since science began researching and influencing structures on a nano-level, new carbon materials have been on the rise, making new solutions possible across a wide variety of applications and providing longevity as well as lightweight construction. The best known of these is carbon nanotubes (CNT), a particularly stable configuration of a hexagonal honeycomb structure along a pipe measuring just a few nanometers. It is a mechanically heavy-duty material, and in theory is around five times more stable than steel and twice as hard as a diamond. The potential of CNT for lightweight construction is obvious, and a range of compound lightweight construction systems with CNT reinforcement are already appearing on the market.

Properties particularly stable carbon configuration // hexagonal honeycomb structure // CNT five times more stable than steel // twice as hard as diamond // huge increase in stability // electrically conductive // electromagnetic shielding

Sustainability aspects saving resources through efficient use of materials in structures // energy-saving through weight reduction

PRODUCTS

Particle foams with CNT admixture

CNT admixtures in **particle foams** (EPS, EPP) hugely increase the robustness of **foam materials.** These offer improved possibilities for use in body armor, motorbike helmets, and bumpers. Developers are currently testing granulate materials with 10% CNT added before the foam process for the production of particularly fine-celled structures.

CNT-modified polymer composite

FutureCarbon is working on improving carbon-fiber-reinforced synthetic building components for use in space exploration as part of its Carbo Space project. CNT-modified polymer composite materials are being produced which can withstand the extreme conditions of space travel and make ultra-lightweight construction solutions possible. The addition of carbon tubes gives polymer compounds improved **electrical conductivity** and **electromagnetic shielding functions.** When it comes to integrating carbon tubes, the qualities of the matrix material are particularly crucial. Here, researchers have managed fundamentally to improve the mechanical properties of the reactive resin on the basis of epoxide or cyanate ester resin. Its break and shearing resistance can be improved by around 50%, with considerably higher use temperatures.

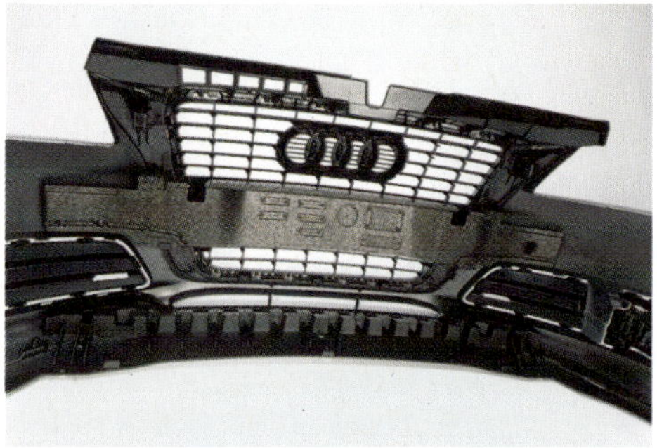

Bumpers made of particle foam with CNT admixture (Source: Ruch Novaplast)

CNT-reinforced aluminum

Zentallium® is a CNT-reinforced aluminum developed for the requirements of the aviation and automobile industries. It is a compound material made using an aluminum base material and CNT reinforcement. The core of the material consists of a **nanostructured aluminum matrix,** in which the **nanotubes** are embedded. The material boasts levels of stability that exceed those of stainless steels and match those of structural steels. Zentallium® is being used, among other things, for designing lightweight constructions for **electromobility.** Most recently, it was successfully tested in a model helicopter (**RC helicopter**).

Zentallium® – CNT-reinforced aluminum (Source: Zoz Group)

RC helicopter with Zentallium® (Source: Zoz Group)

Properties high strength relative to its mass // similar qualities to Kevlar // strong tendency to cross-link // extremely large surface area // extremely reactive // biodegradable molded parts possible // barriers against oxygen and steam

Sustainability aspects based on organic waste // increase in material efficiency

NANO-CELLULOSE

Cellulose is contained in the cell walls of virtually all plant structures as a biopolymer. Cellulose fibers are very strong and so lend themselves to bio-based fiber reinforcement of carbon compound materials. Researchers in the USA, Brazil, and Switzerland have managed to produce cellulose fibers in nano-dimensions and to isolate them as a powder.

MATERIAL CONCEPT AND PROPERTIES

Nano-cellulose fibers have a diameter of less than 100 nanometers and a length of less than a micrometer. The individual fibers are extremely strong relative to their mass and build strong cross-links among themselves. This means they have a very large surface area, which makes them extremely reactive. **Nano-cellulose** forms physical-chemical links with inorganic, organic, and polymer materials, which gives it similar qualities to **Kevlar** as a reinforcement material and makes it particularly interesting for use in composite materials. Just one kilogram of nano-fibers would be sufficient to produce 100 kg of compound synthetic with a significantly reduced weight. Scientists from Empa, a research institute at the Swiss Federal Institute of Technology (ETH) in Zurich, were able successfully to extract nano-cellulose from wood, while researchers at the Universidade Estadual Paulista (UNESP) in Brazil used **pineapple** leaves and the shrub of the **banana tree** to produce a talc-like powder.

APPLICATION

Nano-cellulose powder lends itself to the production of strong polymer compounds with similar levels of stability to those of metal components for the automobile industry as much as to membrane or filter materials in biomedicine. Above all, when combined with biopolymers, it permits the creation of strong yet biologically degradable molded components. Nano-cellulose can also be used to improve the mechanical qualities of wood and paste materials. In the form of **nanoporous bio-foams,** it can be a substitute for conventional insulating materials. Pressed into a thick paper, a nanofiber network with distributed clay particles can be used as a barrier layer for **oxygen** or **steam** in composite packaging and act as a substitute for the aluminum currently being used.

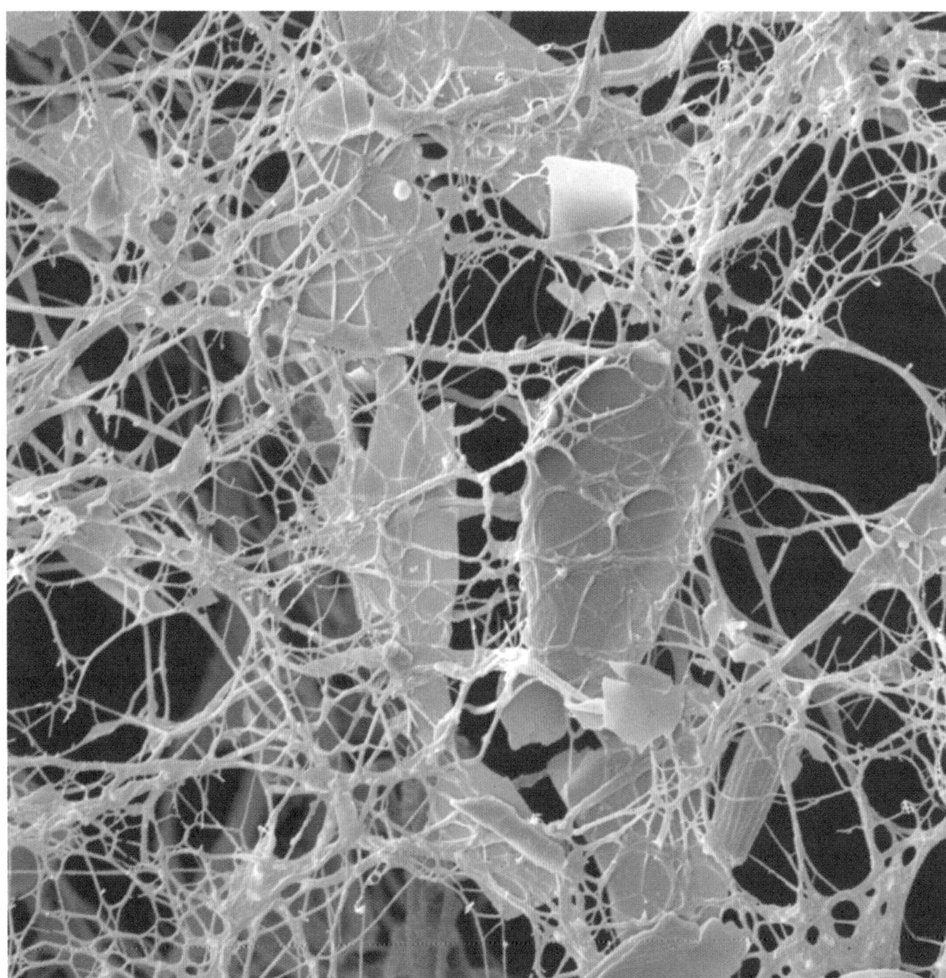

Nanofiber network with distributed clay particles in a shot from a scanning electron microscope (Source: Empa)

Using bioplastics and recycled materials as a basis for the production of foams is gaining in importance. Used in packaging, construction materials, or the automobile industries, they combine qualities such as material efficiency and environmental compatibility with technical properties such as dimensional stability and insulation. Thus, a number of interesting foams made from renewable raw materials and residual plastics have appeared on the market and are finding their way into practical usage.

Properties high level of rigidity // density between 15 and 80 kg/m³ // very good damping // similar to polystyrene foam // industrially compostable // micro-cavities through the injection of carbon dioxide // closed and open-celled structures

Sustainability aspects based on renewable raw materials // increase in material efficiency // biodegradability with the use of a bioplastic

BIO-FOAMS

PRODUCTS

Bio-based particle foams

One of the pioneers of this development is Fraunhofer ICT in Pfinztal, Germany. Here, researchers have recently been able to produce **particle foams** using **polylactic acid (PLA)** and various cellulose bioplastics. Particle foams are **foam beads** of thermoplastic synthetics, which can be further processed with the use of steam to create molded parts for lightweight construction, packaging, or technical construction elements. Their density is usually between 15 and 80 kg per cubic meter, and they are also characterized by very good damping properties.

Extrusion-foamed biopolymers

Alongside particle foams, foamed sheet materials can be produced from bio-synthetics through extrusion. For this purpose, **twin-screw extruders** are used, which plasticize the polymer by mixing in additives, then mix it with liquid and environmentally friendly propellants, and foam it by means of the extrusion process. Extrusion-foamed biopolymers usually boast densities up to a value of 200 kg per cubic meter.

BioFoam®

BioFoam® comes from Dutch producer Synbra Technology and is a **hard foam** made from **PLA.** It has similar properties (compressive strength, insulation, cell structure) to **expanded polystyrene** (EPS) and is made from **cornstarch** or **sugar.** The foam is industrially compostable and can be used for packaging material and insulation.

MicroGREEN Ad-Air

American firm MicroGREEN has developed a technology for the integration of **carbon dioxide** in synthetic foils and polymer sheet materials and has commercialized this for **PET** made from recycled bottles and the bioplastic PLA. The incorporation of this greenhouse gas produces micro-cavities, which expand the base material in length and breadth by 50%, while the material thickness doubles. At the same time, the relative material density is reduced by 80% and the temperature stability is increased. PET treated with Ad-Air can withstand temperatures of up to 300 °C. Also possible are **open-cell structures,** which are currently the subject of research.

Molded parts made of bio-based particle foams (Source: Fraunhofer ICT)

Ad-Air technology for the production of expanded PET (Source: MicroGREEN)

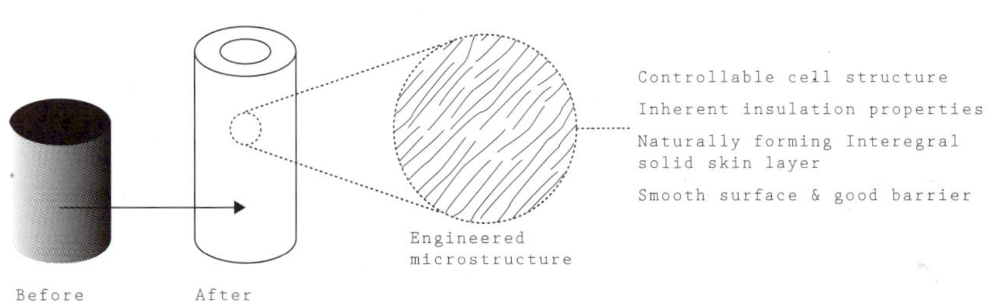

BioFoam® packaging (Source: Synbra Technology)

Properties high level of stability combined with enormous strength // very thin walls // good damping // inflated pneumatic structure with integrated fibers // lightweight construction

Sustainability aspects Minimal use of materials

BIOMIMETIC LIGHTWEIGHT CONSTRUCTION

Over millions of years, nature has perfected organic structures with regard to their mechanical stability in relation to use of materials. One example is human bones with their spongiform cell structure, or diatoms, whose ribbed and radiating honeycomb structure offers a high level of stability combined with enormous strength. The potential for application in technology is tremendous and so numerous projects are under way in which bionic scientists are trying to reproduce such designs, translating lightweight structures from nature into industrial applications.

Technical plant stems

The **stems** of **sweet grasses** are always hollow and have very thin walls, but boast extremely high levels of stability and notable damping properties. These qualities are due to hollow channels that run through the stem walls. Scientists at the Institute of Textile Technology and Process Engineering Denkendorf and biologists from Freiburg have successfully replicated the material-efficient construction principle of plant stems (in particular **horsetail reed** and **bamboo**). These patented **"technical plants"** are produced by means of a **braided pultrusion process.** In this process, plant fibers are interwoven, soaked in resin, and pulled through a heated mold. This hardens the fibers and the gradually produced hollow structure gets its inner geometry. Technical plant stems boast a particularly high resistance to buckling and bending and offer potential for application in the fitting of conduits in vehicle construction or in tubular structures for sport and architecture.

Pneumatic structures

There are countless examples of an inflated **pneumatic structure** as a construction principle to be found in nature. Almost all plant structures function according to this principle, because the stability of a tissue is created by cells under pressure, whose cell walls are also protected from tensile and bending stresses by integrated fibers. At Empa in Zurich, bionic scientists have developed a structure for a **mesh-reinforced pneumatic structure** on the basis of this principle, with which highly stable lightweight structures can be realized in a straightforward way for use in construction and in aviation. A prime example of the inflated pneumatic structure is the roof of a multistory car park in Montreux. For military applications, it offers the possibility of rapid assembly and dismantling. The development is being marketed under the name **Tensairity®.**

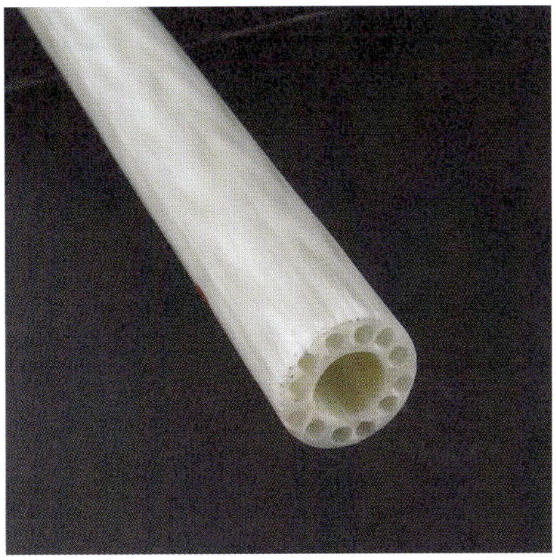

Technical plant stems
(Source: Institute of Textile Technology and Process Engineering Denkendorf)

Tensairity® kite
(Source: Empa)

Multistory parking garage in Montreux with Tensairity® pneumatic structure
(Source: Empa, design: Luscher Architectes SA & Airlight Ltd.)

Textile structures filled with air are familiar to us from their application in sport and leisure activities. However, they also offer numerous advantages for technical constructions, aviation and architecture, and are the focus of a whole range of developments with an eye on harnessing their potential for lightweight construction.

Properties air cushion system for weight-saving // sufficient stability for supporting structures // lightweight transportation // air dome with pressure lock // compensation for pressure changes through flexibility of the material

Sustainability aspects minimum use of resources // energy-saving when used in aviation // reduced assembly and transport costs

PNEUMATIC TEXTILES

Pneumatic comfort system

Swiss textile manufacturer Lantal, for example, has developed an **air-cushioning system** that is protected by various textile layers. It saves up to 3 kg in weight per business-class seat. Passengers can adjust their seats for comfort individually and with infinite variability.

Self-supporting structure with air-filled tubes

In architecture, **gridshells** are among **surface load-bearing structures** with which large-scale roof structures can be realized without supports and with minimum use of materials. To test the potential for use of **pneumatic textiles** in this area, the Institute for Lightweight Structures and Conceptual Design (ILEK) at the University of Stuttgart has developed a gridshell with integrated **air tubes.** The **air-supported membrane** makes assembly easy and keeps the transport weight down, so offering a good solution for mobile architecture for events and exhibitions.

Air dome
(Source: Paranet Germany)

Air-supported architecture

Pneumatic air-supported structures are principally used in architecture for the construction of lightweight roofs, temporary halls for exhibitions and sports events, or to cover **biogas plants.** For the erection of an **air dome,** the textile membrane structure is attached to the ground by means of a **cable mesh,** for example, and raised under low upward pressure from inside. Since lightweight inner pressure is essential for the stability of the construction, air domes are accessed via an **air lock.**

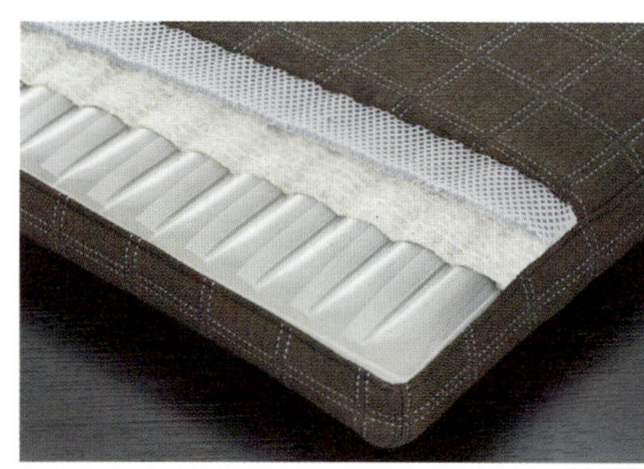

Structure of the pneumatic comfort system
(Source: Lantal)

Architectural structure with the use of air-filled tubes
(Source: ILEK, University of Stuttgart)

For biogas plants, nowadays **double-membrane air-supported roofs** are often used, which balance changes in volume by lowering or increasing the temperature in the interior and adjusting the pressure. The external layer of the double membrane protects against weather and UV radiation, while the inner layer functions as the actual **gas storage.** An inflated air layer achieves stability and has an insulating effect. Two-layer solutions are notably more durable than their one-layered counterparts.

Double-membrane air-supported structure for biogas plants (Source: Ökobit)

AEROGRAPHITE

In the summer of 2012, a new material based on carbon was introduced which, in its lightness, far surpassed the previous world record-holder, a nickel material.

Properties based on carbon // lightweight construction world record // 99.99% trapped air // light-absorbing // water-repellent // compressible // electrically conductive // high resistance to vibrations

Sustainability aspects clear weight advantage // energy-efficient production

MATERIAL CONCEPT AND PROPERTIES

Aerographite is black, absorbs light radiation almost completely, and is electrically conductive. It was developed by scientists at the Hamburg University of Technology and the University of Kiel and has a density of just 0.2 milligrams per cubic centimeter. This means it weighs 75 times less than the polymer foam **polystyrene.** Aerographite consists of a network of porous **carbon tubes** at nano-level, which make the material stable and able to withstand tensile and pressure stresses despite its low mass and 99.99% trapped air. Its structure offers advantages in terms of stability over and above the likewise ultralight but fragile **aerogels.** It can be compressed like a **sponge** and reduced to 95% of its original size.

APPLICATION

When it comes to electric conductivity, researchers are expecting the first application possibilities to emerge for particularly lightweight battery systems in the context of **electromobility.** With aerographite, plastics can also be made conductive for polymer-electronic applications. For satellite and aviation electronics, there is great potential due to the material's high resistance to vibrations. Aerographite can also be used for purifying water or air for incubators or ventilators. Although the production costs of the sponge structure are comparatively low, it will be at least another 10 years before it is put to use in industry.

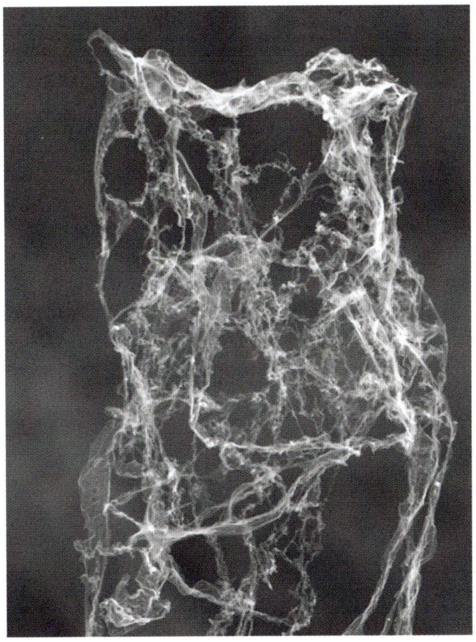

Open carbon tubes form a fine network (Source: Hamburg University of Technology)

Hydrophobic qualities of aerographite (Source: CAU)

MULTIFUNCTIONAL MATERIALS

05

A space that responds to changes in ambient climate, walls and tiles that change color or structure when the temperature rises, furniture surfaces and fabrics that exude a pleasant scent: for some years now designers have been busy testing the potential for so-called smart materials. A whole raft of product concepts have already been made public, but to date product design seems to have eschewed intelligent materials – despite the fact that scientists have developed a wide range of new ones.

The most obvious examples of intelligent surfaces include paints that are temperature or water-sensitive, triggering changes in the color or transparency of surfaces, foils, or fabrics. This affords designers an additional creative level, and hardly surprisingly interior designers are increasingly interested in materials that enhance the ambient climate and eliminate unpleasant odors or toxic substances. New ways of integrating electronic components and heating technologies equip materials with additional functions that can help reduce the inputs required to manufacture the relevant products. Among the highlights here are new functional materials with shape-changing properties, such as electroactive elastomers or shape-memory materials, as well as those whose mutability changes depending on outside forces or electrical fields. In particular, research has been intensified worldwide in recent years in the areas of metamaterials, i.e., materials with qualities that are not encountered in the same guise in nature.

125

MULTIFUNCTIONAL MATERIALS

Properties color change under the influence of UV light, heat, moisture, and magnetism // color change adjustable // processing time of impression materials shown

Sustainability aspects signal function with zero energy consumption // no need for sensors and other electric components

COLOR-CHANGING MATERIALS AND SURFACES

With the emergence of water-sensitive, thermochromic and pH-responsive surfaces, the opportunities for working with color-changing materials have increased appreciably for designers and architects. It is now becoming possible to create spaces and objects that transform with increasing temperature or changing moisture conditions.

PRODUCTS

Thermochromic ceramics

Kitchen and bathroom designers hoping to work with **color-changing glass and ceramic tiles** can find what they are looking for at the Californian supplier **Moving Color.** The tiles produced by this company are black at room temperature and change color across the entire spectrum as they heat up from the shower water or when a fire is lit or heater switched on.

Temperature-sensitive colors

We can now define the reaction time of heat-sensitive colors with great precision, i.e., when the color changes on reaching a certain temperature. **Temperature-sensitive colors** can, for instance, assume security-related signal functions as under normal climatic conditions the color remains constant. **Thermolock® colors** can be used to indicate breaks in the cool chain in logistic processes.

Photochromic inks

These inks react to **ultraviolet light.** They are generally used as additive colors, for example, to produce color variations in strong sunlight in print

Heat-sensitive tiles
(Source: Moving Color)

Reaction times of the various Thermolock® colors (Source: Matsui Color)

	Termperature at which color begins to change
Type	
Thermolock® 11	-2 °C
Thermolock® 23	9 °C
Thermolock® 39	22 °C
Thermolock® 48	34 °C
Thermolock® 72	45 °C
Thermolock® 79	45 °C

Essence Magnetic polish
(Photo: Diana Drewes)

Water-sensitive clothing
(Design: SquidLondon)

images for textiles or stickers. They commonly change from being transparent to having a slight coloring. **Photochromic inks** are also used for security applications.

Color-changing impression material

The most common irreversible impression material that has elastic properties for use in dentistry is **alginate.** It is obtained from brown algae and has no negative effects on the body. This also makes it suitable for artistic **body casts,** producing high accuracy results down to the pores. As the material hardens quickly and becomes useless when improperly processed, there are so-called **color-changing alginates** on the market. These contain an additive that reacts to the changing **pH value** during the hardening process. The material is prepared in such a way that it changes color to indicate all the relevant steps in the process, including the mixing, processing, and setting phases.

Water-sensitive colors

In addition to photochromic and heat-sensitive pigments, there are also those on the market that react to moisture and, for instance, change from opaque white to transparent. They are used, among other things, in umbrellas, jackets, towels, shower accessories, and facades. On drying, the pigments return to their original opaque white.

Magnetic lacquer

The nail polish industry is a pioneer in the development of variable solutions. A new nail polish contains magnetic particles that under the influence of a **magnetic field** align themselves in such a way that they form a **diamond pattern.** The bottle cap contains a magnet to produce the effect. The more layers of the **magnetic polish** applied, the darker the diamonds become.

Water-sensitive umbrella
(Design: SquidLondon)

ANTIGRAFFITI COATINGS

Properties decrease in surface tension // water and oil repellent // temporary or permanent protection

Sustainability aspects increase in durability // reduction in cleaning effort // decrease in water consumption

The annual cost of removing illegal graffiti on public and private facades is estimated to be around €500 million in Germany alone. The procedure is generally complex, and the paint cannot always be completely removed. antigraffiti coatings can help here. They can be temporarily applied even to porous surfaces and reduce the adhesion of paint and dirt.

MATERIAL CONCEPT AND PROPERTIES

Sacrificial protective coatings

Temporary protective coatings are as a rule based on **wax solutions,** which are applied as a **sacrificial layer** and protect walls or facades for a certain period of time from the application of paint and, being water-repellent, counter the effects of the environment and moisture. A **high-pressure cleaning system** is used to clean the surfaces, which also removes the **protective coating.** Anti-graffiti wax is widely used and is suitable both for smooth and for porous surfaces. Bio-based waxes are less suited to use as protective coatings.

Semipermanent coatings

Single- or multiple-layer semipermanent systems guarantee protection for around three cleanings with no reduction in effectiveness, are water- and oil-repellent, and can be applied to numerous surfaces. Compared with **permanent systems** they have better optical qualities, as they do not clog pores. However, given that they need to be constantly renewed, they are not suitable for protecting listed structures.

Permanent protective coatings

Special **coating systems** offer permanent protection from the unwanted application of paint. These coatings can be cleaned at least 15 times without any reduction in efficacy, although the use of thick protective coatings considerably lowers the aesthetic quality of porous surfaces. An anti-graffiti aluminum sheet with high resistance to corrosion and weathering was recently launched under the name **CLEARKY 105.** Its effectiveness is based on a three-layer system with **fluoropolymer lacquers.** Impurities can be quickly and easily removed with a high-pressure cleaner or scraper. The coating does not impair processing steps such as punching, roll forming, or trimming. ThyssenKrupp offers a permanent protective system marketed under the name **PLADUR® L,** which takes the form of a **steel strip.** The protective components are a **clear lacquer** and a special additive.

Permanent anti-graffiti protection for aluminum facades (Source: Novelis)

FUNCTIONAL ORGANOSILANES

Anti-adhesive coatings are highly significant in architecture as they extend the period between expensive cleaning cycles. They are mostly based on functional organosilanes, which also protect against corrosion on metallic surfaces.

Properties anti-adhesive coating for metal and glass // water-repellent properties // very good adhesive properties // increase in the strength of bonds

Sustainability aspects reduction in corrosion // longer cleaning cycles // reduction in fuel consumption owing to lower rolling resistance // noise reducing

MATERIAL CONCEPT AND PROPERTIES

Silanes are a group of substances consisting of a **silicon structure** and **hydrogen**. By means of a process known as **silanization,** they are chemically bonded to material surfaces to increase their functionality. The attainable functionalization depends, among other things, on the type of the silane. Organosilanes have very good adhesive properties. As a **bonding agent** they increase the strength of bonds between polymeric materials and metals, glass or mineral materials. In particular, they improve the tensile properties of materials under the influence of moisture.

Effectiveness of silane bonding agents on various surfaces:

Graphite:	none
Nickel and zinc:	moderate
Iron and steel:	moderate to good
Aluminum:	good to excellent
Glass:	excellent

APPLICATION

Silane-based bonding agents are key components in adhesives and sealants for the construction industry, transportation, and aerospace, where they replace complicated welded joints and bolted connections. In easy-to-clean **hydrophobic coatings** they are used on facades as well as in shower cubicles and other hygienic areas. Moreover, their adhesive strength when in contact with moisture can be used to positive effect in dentistry. Functional organosilanes developed especially for tires serve to lower **rolling resistance,** which leads to a reduction in noise as well as fuel consumption.

Properties bacteriostatic and anti-inflammatory // antibacterial // antifouling effect // release of active agents

Sustainability aspects materials based on natural substances biodegradable // positive effect on wellbeing

ANTIBACTERIAL SURFACES AND FIBERS

There are various methods and additives for creating antibacterial surfaces and fibers. Silver and copper are traditionally used as the bacteriostatic materials to make door handles, cutlery, and coins. The Ancient Egyptians were probably the first to recognize this function. Indeed, there is a long history of integrating silver threads in textiles in medical technology. In 2012, however, scientists at the University of Duisburg-Essen proved that silver ions can also have a negative impact on human tissue. Alternatives would be materials like zinc, magnesium, or iron to promote the healing of burns, or natural fiber materials with antibacterial properties such as milk and soy protein fibers, corn fibers, and algae agents.

PRODUCTS

SeaCell® fibers with algae

Using a production method developed by the Thuringian Institute of Textile and Plastics Research (TITK), smartfiber AG integrated natural additives into **cellulose fibers** thus furnishing them with skin-protection and antibacterial properties. The anti-inflammatory qualities of algae come from sea salts and minerals, which the plant fibers absorb from the sea water. **SeaCell® fibers,** which are completely biodegradable, contain above all **brown algae** (also known as **knotted kelp** or ascophyllum nodosum). Medical studies have shown that knotted kelp includes the active agent **fucoidan,** which slows the growth of tumor cells.

Cellulose fibers with the trace element zinc

As a trace element, **zinc** can also have antibacterial effects in cellulose fibers. Moreover, it has a regenerative effect on the skin, enhancing the healing process of inflammation and wounds. This is triggered by body moisture, which enables active exchange between skin and fiber. This both disrupts the metabolism of the bacteria and prevents the development of unpleasant odors.

Finishes with bactericidal nanoparticles

A major problem in seafaring is the fouling of ships' hulls by barnacles, algae, or all kinds of shells. Their presence increases resistance in the water, impairs handling of the ship, and leads to an increased energy requirement (up to 28% higher fuel consumption). Researchers at Bremen University of Applied Sciences spent years studying shark skin and then created an **antifouling spray.** Now scientists at Johannes Gutenberg University Mainz have discovered that the biofouling process can be prevented using **vanadium pentoxide** nanoparticles. They found an analogy in enzymes in brown and red algae, which are capable of producing halogen compounds with a **biocidal effect**, and in this way protect themselves against microbial infestations.

Textiles used as the base for active agents

After seven years of research, Swiss company Schoeller® Textiles presented a technology at Techtextil in 2011 named "iLoad," which can temporarily store therapeutic agents and release them over a certain period of time (**desorption time**). The core is the so-called **donor layer** that encases the textile fibers. This layer can be loaded with the desired active agent in a few hours using the rinse program on a washing machine. As the textile is worn, friction, heat, moisture, and sweat cause the release of the agent. Residues are removed the next time it is washed. The **iLoad textile** can then be reloaded with an active agent.

SweetSkin

A fiber for the sports and medical sectors has been launched on the Spanish market. It contains anti-inflammatory aloe vera gel housed in microcapsules which is released in small doses upon contact.

SeaCell® fibers with the active agent algae (Source: smartfiber AG)

Functional principle of the textile iLoad releasing the active agent
(Source: Schoeller Textil)

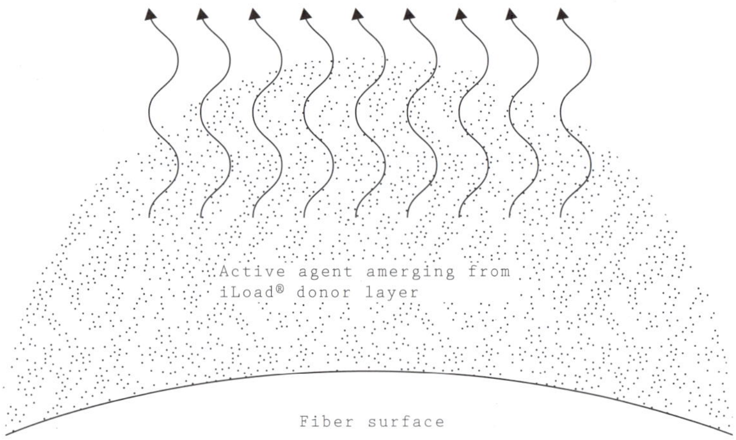

Boat's hull with barnacles
(Source: Bremen University of Applied Sciences)

Properties possibility to wash textiles, bleach paper, stain leather // enzymatic discoloration // antimicrobial wood surfaces // improvement of adhesion // biological derusting

Sustainability aspects biodegradable // substitution of environmentally harmful substances and acids

FUNCTIONAL ENZYMES

Technical enzymes with functional properties are becoming increasingly widespread. In laundry detergents for textiles, paper bleaching, and leather staining they replace the chemicals used hitherto, which are damaging to the environment.

MATERIAL CONCEPT AND PROPERTIES

Enzymatic textile finishing

In the search for more environmentally friendly processing methods for materials, scientists have also found solutions for textile finishing among bio-based processes. For whereas jeans have hitherto been washed with **pumice stones** to lend them the worn aesthetic favored by many buyers, a company from Darmstadt has now replaced that archaic finishing process with an enzymatic one. It uses cellulases and laccases to treat the fabric. While **cellulases** break down the dyed outer layer of cotton fibers, **laccases** oxidize the indigo dye. The fibers lose some of their color, without suffering mechanical wear.

Enzymatic wood functionalization

Certain types of fungi that attack wood, such as **white rot fungi,** excrete enzymes that can be used to equip wood surfaces with additional functions. A team of researchers at EMPA in Zurich is working on the targeted use of the enzyme **laccase** to lend wood an **antimicrobial surface.** This would have benefits above all in wood applications in public buildings (hospitals, schools, etc.) and transportation. Moreover, laccase enzymes could be used to improve wood's tendency to adhere to other materials and enable stronger adhesive bonds. In this way, wood with poor adhesive qualities, such as beech or larch, can be bonded better using glue.

Biological derusting with enzymes (Source: ASA Spezialenzyme)

Biological derusting agents

In a joint project with Peter M. Kunz from Mannheim University of Applied Sciences, ASA Spezialenzyme GmbH has optimized a strain of bacteria that excretes iron-bonding substances and can be used for the biological derusting of sheet steel. The bio-derusting agent based on siderophores can thus replace the acid baths commonly used. After use, the derusting solution can serve as a substrate for plants with iron-deficiency disease. ASA has developed an efficient technique for easily obtaining siderophores.

Principle of enzymatic wood functionalization (Source: Empa)

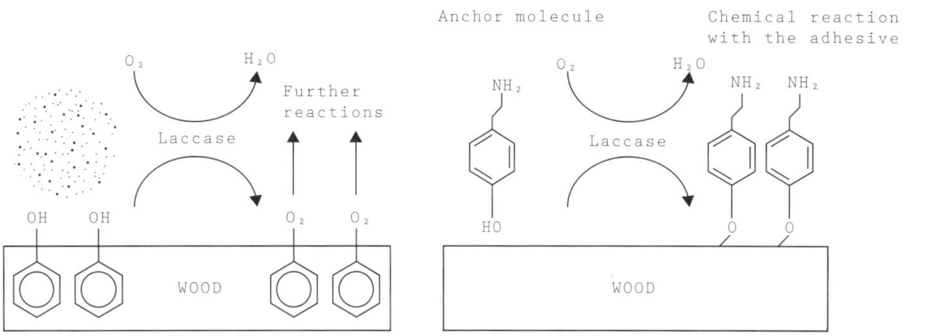

Properties dirt dissolving // water-repellent // UV light blocker // degradation of odors and harmful substances

Sustainability aspects degradation of environmental pollutants in the air // reduction in cleaning

AIR-PURIFYING SURFACES

Scientists have known for some time that titanium dioxide has an air-purifying quality on the nanoscale, which frees particles in the air that can be detrimental to health. With several smaller firms having established the first such products on the market – air-purifying coatings for building products – in recent years the number of product and material innovations with air-purifying properties has risen.

PRODUCTS

Air-purifying ceramics

HT is a durable tile finish with **dirt-repellent** and air-purifying properties. **Nano titanium dioxide** is burned in as a **catalyst** at a high temperature and triggers a reaction between light, oxygen, and air moisture. The **photocatalytic reaction** can be repeated as often as desired, with no reduction in effect. Scientific studies show that 1,000 square meters of HT facade area purifies the air as effectively as 70 medium-sized deciduous trees. Moreover, HT makes tile surfaces **hydrophilic** and thus extremely low maintenance. Water is not repelled, but spreads out as a thin film over the tile, ensuring the effortless removal of dirt.

Air-purifying cement

Under the brand name **TioCem®** a **cement with photocatalytic properties** was launched that, if exposed to daylight, is able to remove up to 90% of **nitrogen oxides,** aldehydes, benzene, and chlorinated aromatic compounds in the air. The cement offers potential for paving stones, flagstones, road surfaces, and noise barriers, particularly in the context of urban traffic.

Dyes to degrade harmful substances

In the context of the joint project **"HelioClean,"** nanoscientists at the University of Kassel are currently developing mixtures of **dye molecules** and titanium dioxide nanoparticles in order to enable the large-scale use of air-purifying paints in road construction. Emissions of toxic exhaust fumes and nitrogen oxides could be considerably reduced with appropriately treated noise barriers or guide rails along freeways and highways. Exhaust fumes are actually only decomposed when high-energy, ultraviolet light shines on the nanoparticles. The chemists in Kassel are seeking to modify dyes such that the effect is also produced with lower-energy, **long-wave light.**

Interior paint to degrade nitrogen oxide

The photocatalytic sealant **StoPhotosan NOX** is able to decompose dangerous **nitrogen oxides** and ozone in interior spaces. The effect is produced by the catalyst titanium dioxide, as long as the crystals are supplied with energy by means of electromagnetic waves such as light. Another Sto product with an air-purifying effect is **Climasan,** which does not require UV light.

Protectin as air purifier

Natural materials, such as components of sheep's wool, are also capable of breaking down odors and harmful substances to the benefit of the indoor air climate. The effect is down to **reactive amino acid chains** in **sheep's wool proteins,** which can transform **formaldehydes** or odors into harmless substances. The **protein complex protectin** can be used to recreate the effective mechanism in sheep's wool. Laminate flooring specialist **Parador** employs the protein to degrade substances harmful to health in laminate and parquet flooring, integrating it into the long edges of the panels. The mechanism has been successfully tested numerous times, with trials proving full effectiveness over a period of at least 44 years.

Air-purifying ceramics with HT finishing (Source: AGROB BUCHTAL)

Guard rail (Source: Saferoad RRS GmbH)

Functional principle of laminate flooring with air-purifying protectin (Source: Parador)

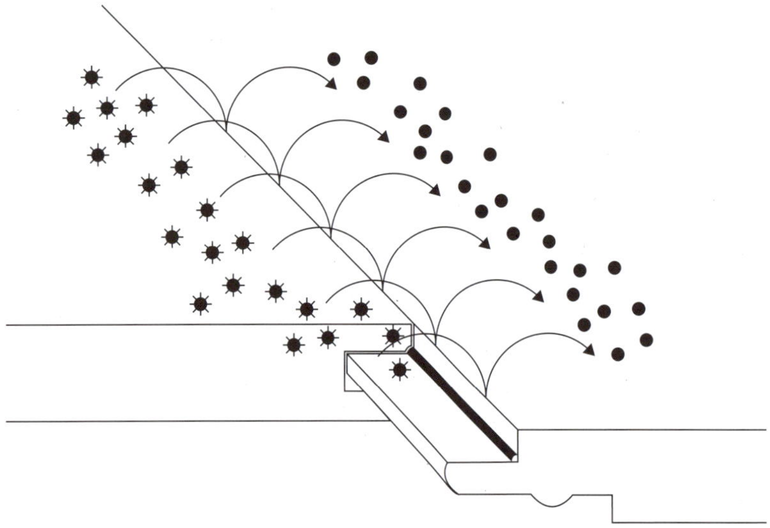

TEXTILE-INTEGRATED ELECTRONICS

Properties electrically conductive // energy storing // energy generating // LED integration // sensor qualities

Sustainability aspects reduction of materials used and number of components

By integrating electronic components, scientists are currently attempting to add numerous functions to textile materials, such as sensor technology, illumination, and energy-generating properties.

PRODUCTS

Conductive fibers

Textile-integrated electronics make use of partially conductive textile structures. To this end, metallized threads are employed, whose electrical conductivity can be specifically modified, for example with a special electro-chemical post-treatment process developed at the TITV Greiz. The advantages of **ELITEX®** threads are their textile properties and ease of use on **textile machines.**

Stretchable circuits

Elastic circuitry is indispensable when using electrical circuits in clothing. Researchers working on the **STELLA** project at the Fraunhofer IZM have developed a solution based on thermoplastic polyurethane **foil** and **meandering conducting paths.** Depending on the geometry and shape of the curve, the scientists have attained elasticity levels of 300%. At the points where flexible meets rigid material, **cable locks** were integrated into the copper to enable smooth connection to electronic components.

Functional LED flex substrate

At Techtextil 2013 in Frankfurt, the TITV Greiz presented the first automatized embroidery of LED-studded **light-emitting textiles**. This innovation is based on the newly developed functional **LED flex substrate: functional sequin devices** (FSD). The sequins used in the fashion industry have been converted so that they can be used as carriers for LEDs or other miniaturized components. It is now economically feasible to produce light-emitting textiles with LEDs for large surfaces as well as small ones.

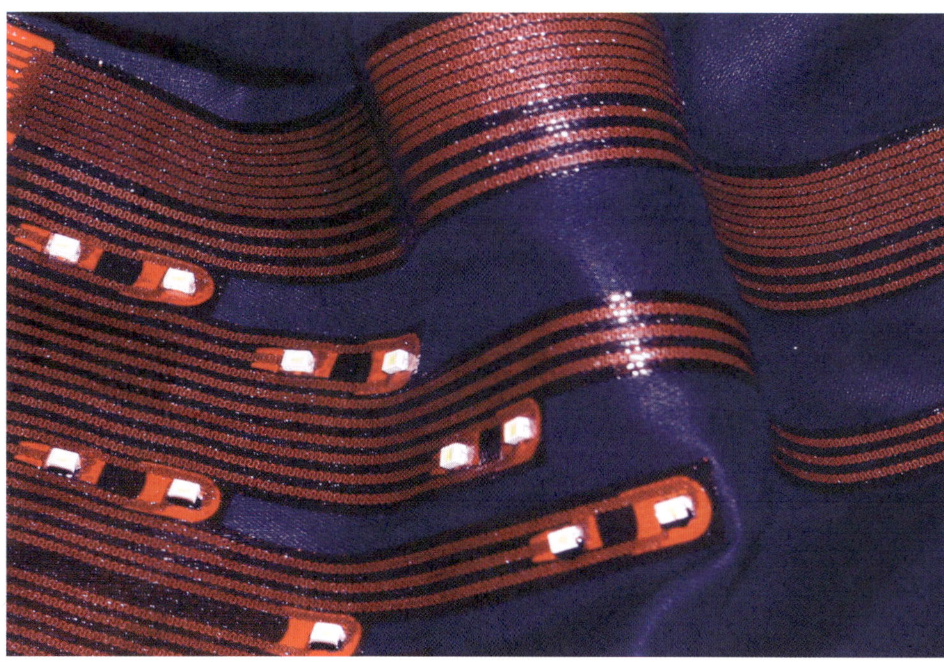

Meandering copper structures on TPU (Photo: Fraunhofer IZM)

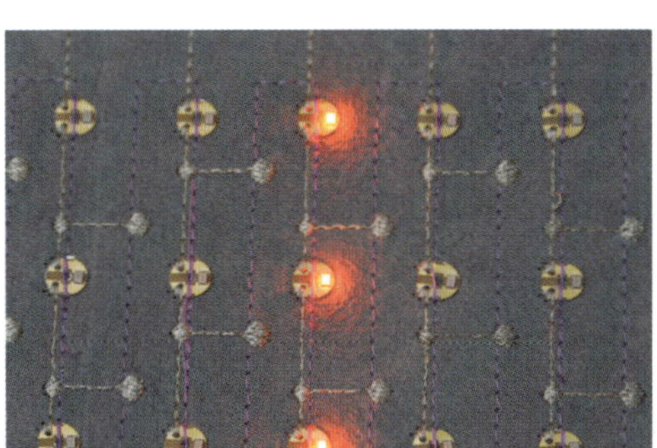

Textile conducting path with embroidered LED sequins (Source: TITV Greiz)

Fabric with integrated LED module (Source: TITV Greiz)

OLED yarn

The project **LUMOLED** was initiated in 2010 with the aim of enabling **textile-based organic light sources.** LED technology based on organic semiconducting materials is considered the next leap forward in the field of lighting. The illumination of flexible substrates is a particular advantage, making OLEDs particularly suitable for use in textiles. Moreover, alternatives are being investigated for either integrating polymeric OLEDs into flexible planar tape yarns, or arranging them on cylindrical fibers.

Textile-integrated sensors and circuits

There are now sensors available on the market that, owing to their functionality and size, can be used in textiles. Named **quantum tunnelling composite** (QTC™), British manufacturer **Peratech** has developed a material that becomes electrically conductive when subjected to pressure. The **textile circuits** have become known in recent years for a series of developments under the name of "smart textiles".

Possible structure for OLED yarn based on cylindrical fibers (Source: STFI Chemnitz)

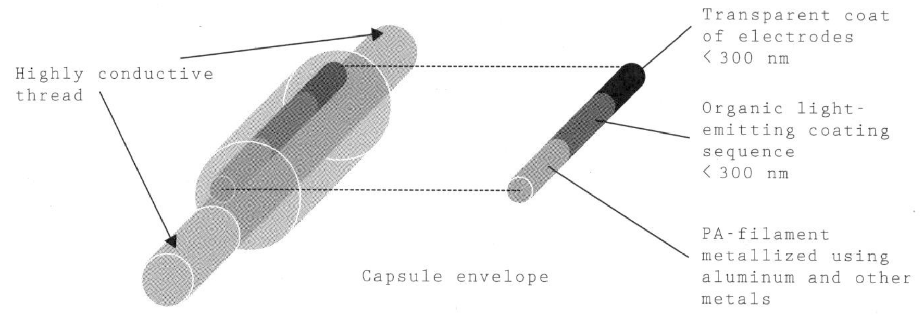

Possible uses of Peratech (Source: Peratech)

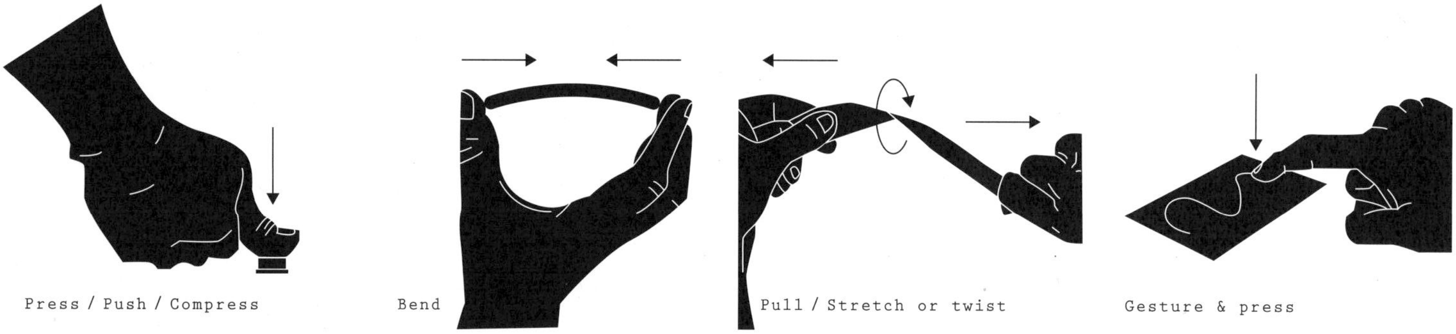

Properties thermal conductivity with carbon fibers and metal threads // temperature regulating with PCM™ // cooling with zeolites // heating system based on reptile skin

Sustainability aspects material-efficient integrated heating and cooling functions // even heat distribution // cooling and heating effect with no energy requirement

HEATING AND COOLING TEXTILES

Textile clothing materials warm us like a second skin, and have done so for centuries. With the migration of the production of classic textiles to Asia, now smart materials with heating and cooling functions are expanding the selection of technical textiles on the European market. Intelligent fibers are starting to be used in entirely new product areas.

MATERIAL CONCEPT AND PROPERTIES

Textiles with carbon fibers

In addition to other outstanding properties, **carbon fibers** are very good thermal conductors. For this reason they are integrated in **heating textiles** to generate heat quickly and efficiently where it is needed. To this end, the heating structure on both sides is charged with low voltage. The heating textile is functionally similar to electric underfloor heating or an **electric blanket** with even heat distribution. The maximum possible temperature is 180 °C.

Textiles with metal threads

Swiss textiles specialists at Sefar have developed technologies with the ETH Zurich to integrate wafer-thin **metal threads** in textiles and use them for heating purposes. They consist of **PET fibers** interwoven with conductive monofilaments. The very short heating phase makes the fabric particularly interesting for sports and outdoor applications. It could be used to heat seats and headrests quickly in the automobile industry. Moreover, the material offers potential for **heated surgical covers.**

PCM™ textiles

PCM™ textiles are special textiles filled with innumerable tiny **wax** microcapsules. These **microcapsules** react to differences in temperature and as they soften they absorb heat from the environment. When body or ambient temperature rises, the capsules store excess heat, and when the temperature falls, they release the heat again. In this way they reduce temperature spikes and improve the normal insulation properties of an item of clothing.

Cooling textiles on the basis of zeolites

In summer 2012, scientists at the Hohenstein Institute in Bönnigheim reported that they had successfully developed a **textile cooling system** on the basis of **zeolites.** The cooling effect requires no electrical energy. It is down to a particular property of zeolites (silicate minerals), namely their ability to adsorb **water vapor** under vacuum and cool down sharply in a very short time owing to **enthalpy of vaporization.** To make use of this property, the researchers developed an airtight hollow textile in the form of a **cooling pad,** which is integrated in a vest and connected to a **metal container under vacuum pressure** containing zeolites. When a valve is opened, the zeolites instantly cool down the water in the **hollow fabric** almost to freezing point. The developers' first idea for the textile's application is to prevent neurological damage in cases of heart failure. It has long been known that cold can protect the brain from the dangers of oxygen deficiency when circulation is poor. Thus the self-sufficient cooling pads are intended to supplement modern mobile **defibrillators** and be used in public buildings and on transportation services by first aiders with no medical knowledge.

Functional principle of PCM™
(Source: Schoeller Textiles)

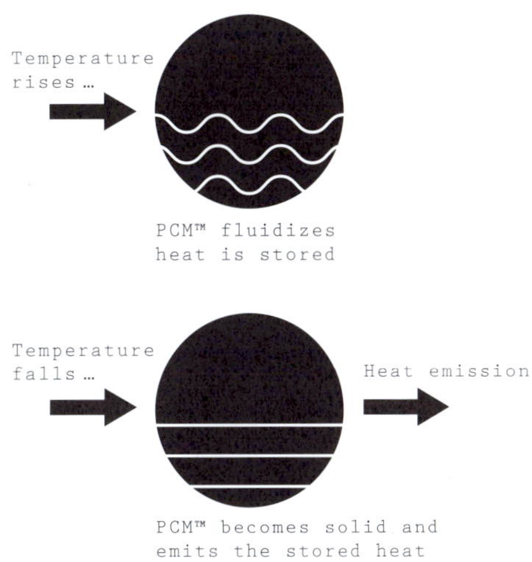

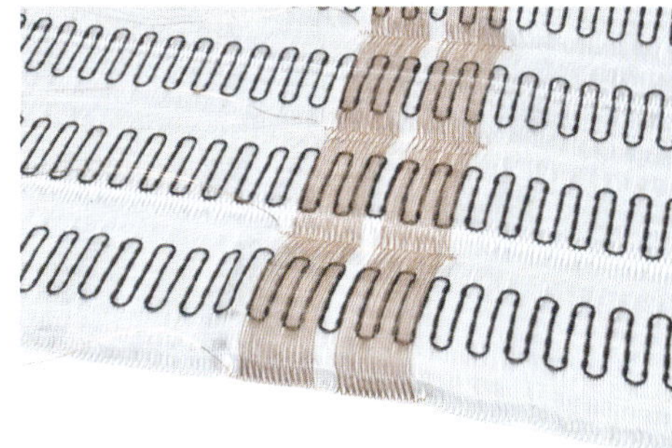

Heating textile with carbon fibers
(Source: Gerster Techtex)

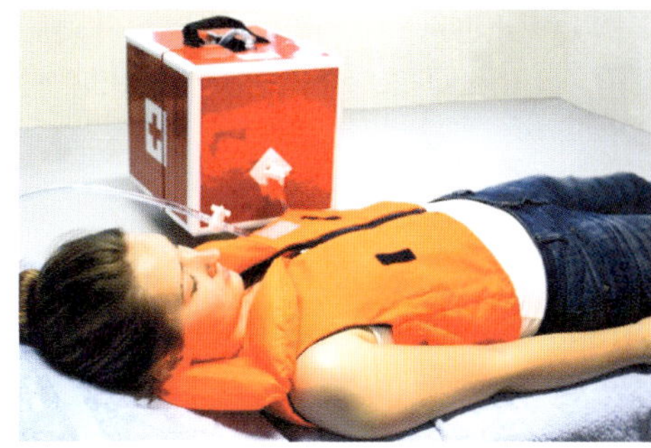

Cooling vest based on zeolites
(Source: Hohenstein Institute)

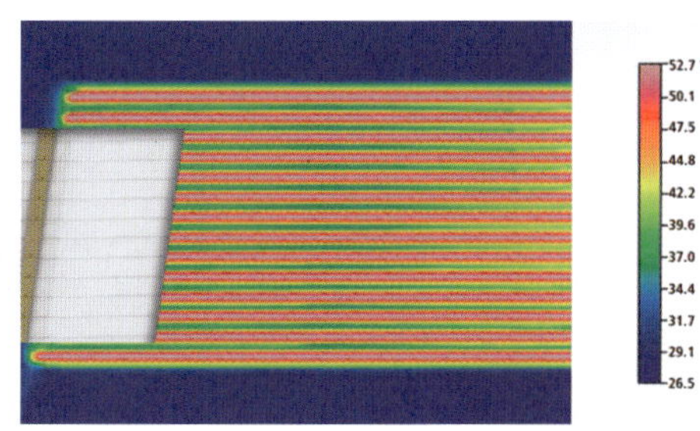

Heated fabric Sefar PowerHeat
(Source: Sefar)

Functionality of the zeolite / water-adsorption technology
(Source: Hohenstein Institute)

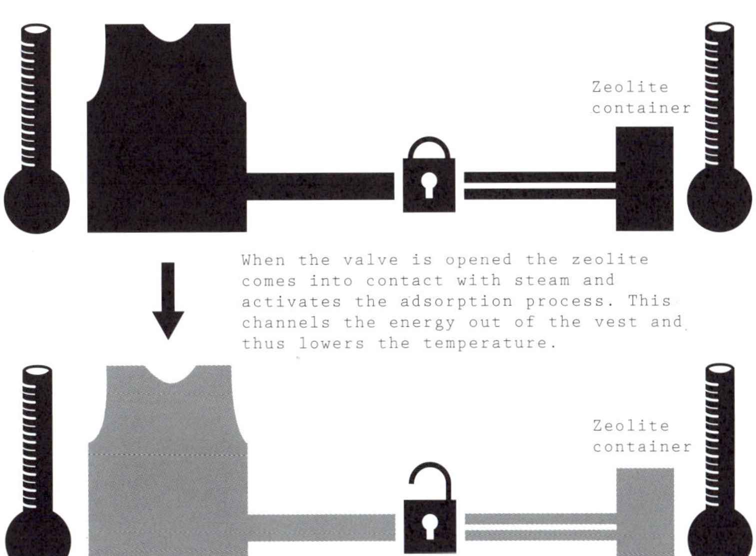

Solar+ textiles

Technology leader Schoeller Textiles has developed this system in analogy to the way the skin of **reptiles** functions, which strongly absorbs natural sunlight. Even thin textiles can keep the body warm, and as such the technology proves its worth in particular in winter sports clothing. The patented system is available for various fabrics.

CNT-HEATED COATING

Properties thermal conductivity: 6,000 W/mK at room temperature // dimensionally stable up to 1,000 °C // safe low voltage // maximum attainable temperature of 500 °C // durable // also suitable for flooring // hard or elastic versions

Sustainability aspects increase material efficiency // considerable potential for lightweight construction // simple assembly // reduce downtime of wind turbines

Carbon nanotubes (CNT) are a prime example of new carbon materials. As additives in polymers, they are capable of conducting heat and transforming a surface coating into a heated surface.

MATERIAL CONCEPT AND PROPERTIES

CNT are mechanically highly stable, chemically resistant, and are outstanding electrical and thermal conductors. Depending on the requirements and desired temperatures, **CNT-heated coating** is available with binder systems on the basis of **acrylate, epoxy,** or **silicone resins,** with a maximum attainable temperature of 500 °C. The coating is durable, making it suitable for use in vehicle interiors, the flooring sector, and bathrooms. Both hard and elastic versions are available, depending on area of use.

APPLICATION

As a coating for **rotor blades,** CNT can be used to de-ice **wind turbines** in the winter, reducing unnecessary downtime. Moreover, with safe low voltage, CNT-heated coatings can be employed as panel heaters for vehicles, underfloor heating for RVs, or wet cells and as tube heaters in medical technology.

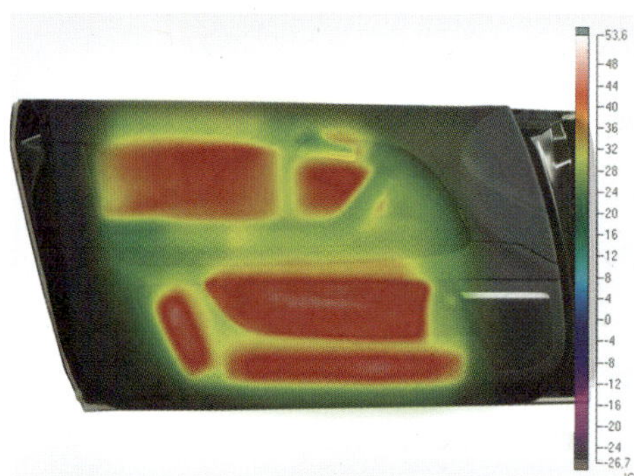

Thermal image of a heated coating in a vehicle interior (Source: FutureCarbon)

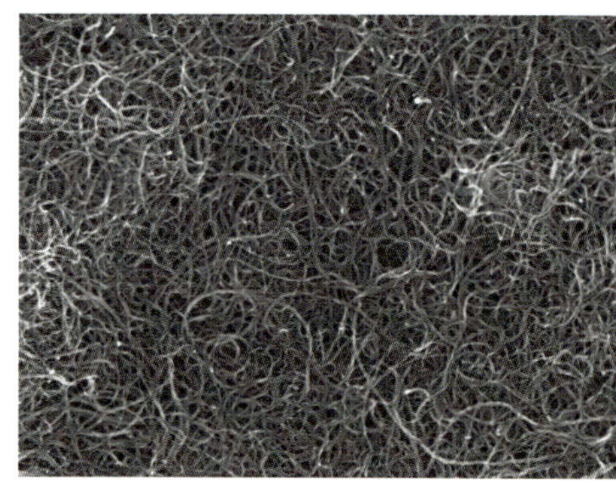

CNT dispersion (Source: FutureCarbon)

GRAPHENE

Properties honeycombed carbon atoms one layer thick // highly conductive // optical transparency // very strong // good rigidity // durable // high electron mobility

Sustainability aspects material-efficient electronic products possible // high storage capacity with low energy consumption

Graphene is one of the most interesting discoveries of recent years. Although scientists have only been researching the material since 2004, it has already been ascribed enormous potential as an alternative to silicon in computer chips and solar cells. Graphene could revolutionize the field of electronics and trigger numerous innovation processes. In 2010, Konstantin Novoselov and Andre Geim were awarded the Nobel Prize in Physics for discovering the material.

MATERIAL CONCEPT AND PROPERTIES

Like **fullerene (Bucky Balls)** or CNT, graphene also consists of **carbon atoms.** It is only one atom thick, has a honeycomb structure, and is an excellent electrical conductor. The single layer is a mere 50 nanometers thick, optically transparent, and extremely flexible. Within the graphene monocrystal, the surface boasts extraordinary mechanical qualities in addition to high strength and good rigidity. If we were to stack layers of graphene, they would theoretically replicate the structure of **graphite.** The material, often referred to as "miracle foil," is superior to silicon in numerous ways. For instance, electric currents can move much faster: **electron mobility** is around 200 times greater than that of silicon.

APPLICATION

Far smaller circuits can be realized with graphene compared with silicon. They could be used to build hard disks with greater storage capacities, and graphene chips with a clock speed of over 100 gigahertz. The transparency of the material may prove advantageous in screens and light-emitting diodes. What still stands in the way of the material's widespread use in diverse applications are the (hitherto lacking) production methods. Which is why, at present, numerous research groups around the world are working on their development. In mid-2012, Samsung was one of the first electronics groups to report that it had successfully built a transistor structure using graphene.

Structural principle of graphene
(Source: Max Planck Institute)

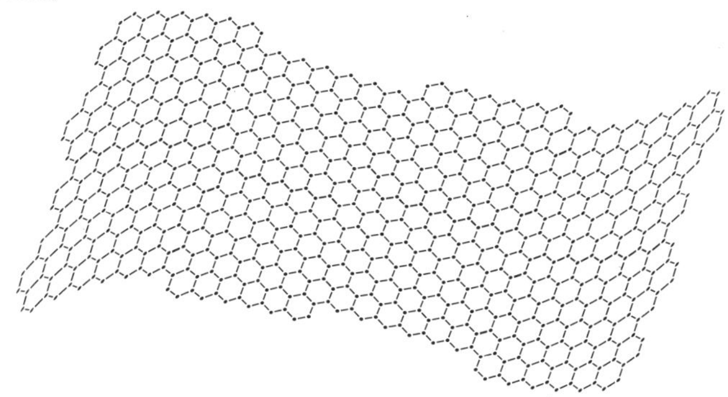

Structural principles of carbon materials

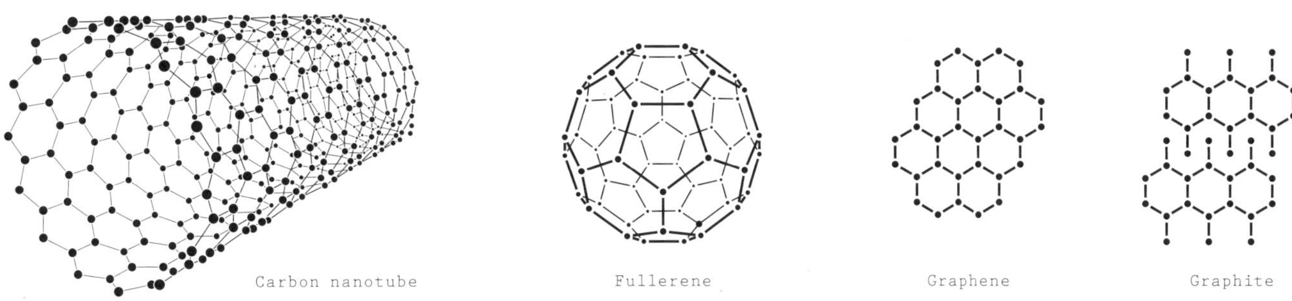

SHIELDING MATERIALS

Properties shield electrical, magnetic and electromagnetic fields // both for high-frequency electromagnetic radiation and low-frequency electrical alternating fields // functionality of cell phones in shielded areas

Sustainability aspects lower electrosmog

Electrical, magnetic, and electromagnetic fields (electrosmog) can have a negative impact on human well-being. To date researchers have been unable to prove this assumption in recognized scientific studies. Yet individual studies repeatedly show up correlations between the number of new cases of illness in a particular area and particularly strong electromagnetic fields. To reduce electrosmog and in the context of biological construction, there are now materials available on the market with shielding functions. Examples include shielding paints, nonwoven materials, films, and fabric or carbon-coated natural gypsum plasterboard.

MATERIAL CONCEPT AND PROPERTIES

Shielding paints

Interior paints with a shielding function generally contain **graphite particles** (carbon) as the conductive element and shield up to 99% of both **high-frequency electromagnetic radiation** (HF) and **low-frequency electrical alternating fields** (LF). Following application, shielding paints are usually covered with a regular colored wall paint. For safety reasons, the coat of paint must be earthed with a **grounding strip**. The functionality of cell phones is limited in rooms painted with shielding paints. When in use, electrical devices must be at least 20 mm away from the wall.

Shielding fabric

The shielding function can also be integrated directly into the exterior of a building in the thermal insulation system with the help of **shielding fabrics.** Normally the insulation is directly applied to the exterior wall and then covered with a reinforcing layer. This consists of an initial layer of plaster – into which a reinforcement fabric is embedded – and exterior rendering. Reinforcement fabric is available on the market with a shielding function for high-frequency electromagnetic radiation and to reduce low-frequency electrical fields.

Shielding clay plaster

Shielding clay plaster is available on the market as a mixture of clay and **graphite fibers.** It is able to conduct heat and protect people in indoor areas from electromagnetic radiation. These qualities complement the already remarkable air-conditioning properties of clay. **EMF shielding clay plasters** are on offer as both base and surface plasters for new buildings and renovation projects. Owing to clay's high heat storage capacity and the good conductivity of graphite, the clay plaster can also be used for **surface heating.**

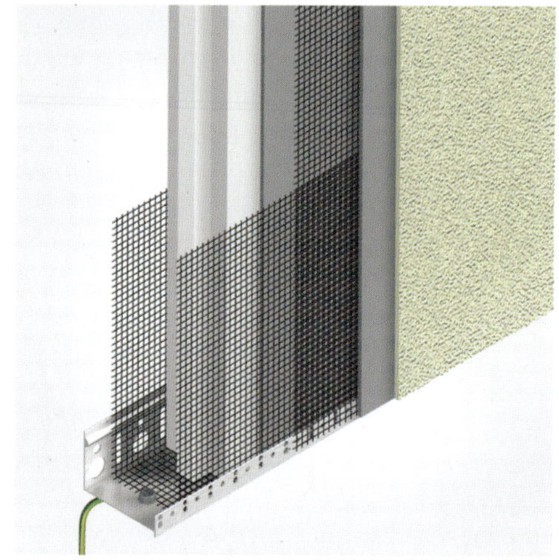

Shielding fabric in a wall structure (Source: Sto AG)

Radiation intensity of selected sources of electrosmog (Source: WDR)

Source	Radiation capacity
Mobile telephone mast at a distance of 60 m	300 mW/m²
WLAN Radio	1,500 mW/m²
WLAN Router	2,000 mW/m²
Cordless telephone	5,000 mW/m²
WLAN Notebook	15,000 mW/m²
Baby monitor	20,000 mW/m²
Microwave oven	150,000 mW/m²

Properties self-healing // scratch resistant owing to microstructuring // sealing // wear protection

Sustainability aspects extend service life // increase material efficiency

SELF-HEALING AND LONG-LASTING MATERIALS

Among intelligent materials are those with self-healing properties. Examples are automotive paints that can reseal scratches themselves, or asphalt surfaces that are able to fill cracks on their own. Not least with regard to airplanes, the potential of self-healing materials is self-evident. Following scientists' presentation of several highly promising solutions for self-healing systems in recent years, the first products are now available to buy.

MATERIAL SOLUTIONS AND APPLICATION

Self-healing polyurethane (PU) paint

One of the first suppliers of self-healing paints based on polyurethane is Bayer MaterialScience. PU paints contain reversible and elastic **hydrogen bonds.** They shift and sever when a scratch forms. The hydrogen bonds then look for new docking sites. Under contact with heat the net reassumes its original structure and the scratches seal up. The **"reflow effect"** offers great potential particularly for the automobile industry.

Self-healing paint under UV light

Swiss and American scientists have developed a polymer coating with embedded metal ions of zinc and also of lanthanum. Incoming **UV light** is absorbed by the metallic component of the so-called **metallo-supramolecular polymers.** As a result the coating heats to over 200 °C within 30 seconds. The scratches melt and seal. This method could be used, for example, to repair bodywork components with local irradiation, without the cost and effort of removing and re-painting them.

Bionic scratch-resistant film

Researchers at Fraunhofer UMSICHT are currently developing a **scratch-resistant film** to protect solar cells and solar-thermal facilities from external influences. They are basing their work on the surfaces of desert roses and **sandfish,** for these show no signs of wear even after sandstorms lasting days at a time. Initial findings indicate that this quality is due to microscopic structures. The scientists are seeking to transfer the **microstructure** onto plastic film using specially structured embossing rollers and nanoparticles.

Self-healing elastomer

The **self-healing elastomer** under development at Fraunhofer UMSICHT is based on the **lactiferous weeping fig.** The plant's milky sap contains capsules filled with **hevein**, which when damaged break open and release proteins. These proteins interconnect with the **latex particles** in the sap and close up the microcrack. The scientists at the Fraunhofer Institute transposed this principle onto plastics technology and integrated adhesive **polyisobutylene** in elastomeric polymers. In so doing they achieved a significant self-healing effect. After 24 hours the tensile extension had repaired itself by 40%.

Self-healing polymer

Reverlink™ is a rubber material consisting of 60% **vegetable oils** and which at the molecular level has **reversible intermolecular bonds.** Cracks and breaks reseal on their own and regain most of their original strength when the two sides are lightly pushed together.

Microscopic image of the surface of sandfish skin (Source: Fraunhofer IFAM)

Function of the reflow effect (Source: Bayer MaterialScience)

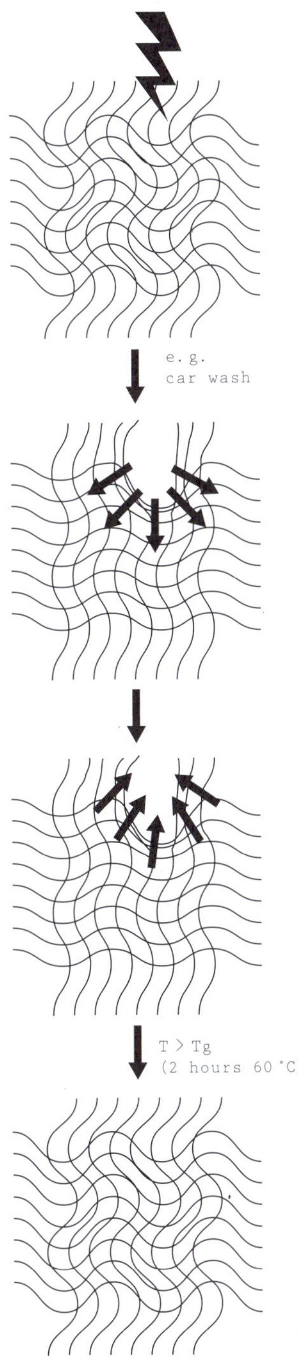

Self-healing hydrogel

In the USA, scientists are working on a **self-healing hydrogel** as a **sealant** for industrial and medical applications. As soon as the jelly-like substance comes into contact with acidic liquids, the numerous short lateral branches of the long molecule chains join up. Cracks and holes close in a very short space of time. The bonds are reversible and break up again in alkaline liquids.

REWITEC wear protection layers

Transmission or **bearing wear** in wind energy plants is one of the greatest challenges in the wind energy sector. As **lubricants** are not always able to permanently protect metal surfaces under all operating conditions, REWITEC has developed a product that improves the surface structure of moving metal parts on a long-term basis. REWITEC's coating technology is based on the modification of the surface structure of moving metal parts by means of the formation of a new, wear-resistant **metal silicate layer.** It increases the resilience of the surface by a factor of 18. The REWITEC wear-protection coating halts damage, such as pitting, and offers protection for wind power gear units and bearings even under extreme environmental conditions.

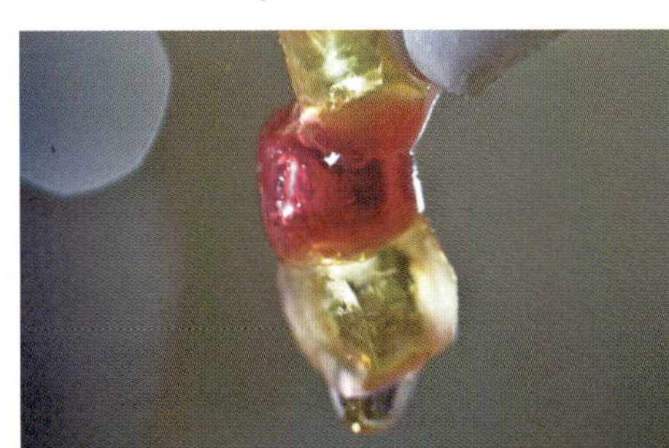

Self-healing hydrogel (Source: San Diego Jacobs School)

Properties high degree of hardness and resistance to corrosion // scratch resistant // elasticity // injection moldable // biocompatible

Sustainability aspects high durability // low processing temperature

METALLIC GLASS

As early as the 1960s, scientists realized that metals can also be produced with an amorphous structure usually found in glass. For a long time, metallic glass could only be produced as coatings albeit with several outstanding properties, but further developments are now enabling the mass production of molded parts by means of injection molding.

MATERIAL CONCEPT AND PROPERTIES

The special qualities of the material are down to the unordered, amorphous arrangement of atoms, which is unusual for metals. Metallic glass is considerably harder and more corrosion resistant than comparable metallic materials. Remarkable properties of metallic glass are the high breaking strength and behavior when struck by another material. Whereas the crystalline atomic structure of conventional metals is squeezed together under external force, in metallic glass it remains intact. The material boasts high **elasticity** alongside very high tensile strength.

Moreover, the **amorphous structure** of metallic glass improves its plasticity under heat. As such it can be injection molded at temperatures of around 300 °C and set in complex formal geometries with extreme precision. This process is accurate to a millionth of a meter. In addition, researchers are working on metallic foams that are 99% air. Their special ability to conduct electrical current enables amorphous metals to react to changes in magnetic fields, which makes them capable of receiving radio waves and opens up potential applications in the field of electronics.

APPLICATION

Metallic glass was first used in the electronics industry in radio receivers or to shield against magnetic fields. To this end, strips were produced by rapid cooling and spraying of the molten glass on a quickly rotating wheel made of a heat-conducting material. In the 1980s, typical metallic glass products included sensors or security tags. Today, metallic glass is available on the market in various alloys (e.g. **zirconium alloy** such as Liquidmetal®), which can be used to make thick and solid components to compete with aluminum or titanium. Scratch-resistant molded parts for the watch and jewelry industry have now also appeared on the market. In the field of sport in particular, the elastic properties of metallic glass can be exploited. In addition to golf clubs, bicycle components, skis, and baseball bats are also now in development. And given that metallic glass based on innovative **magnesium-zinc-calcium alloys** can be biocompatible, it has recently found use in the setting of broken bones.

Behavior of metallic glass when struck by a ball

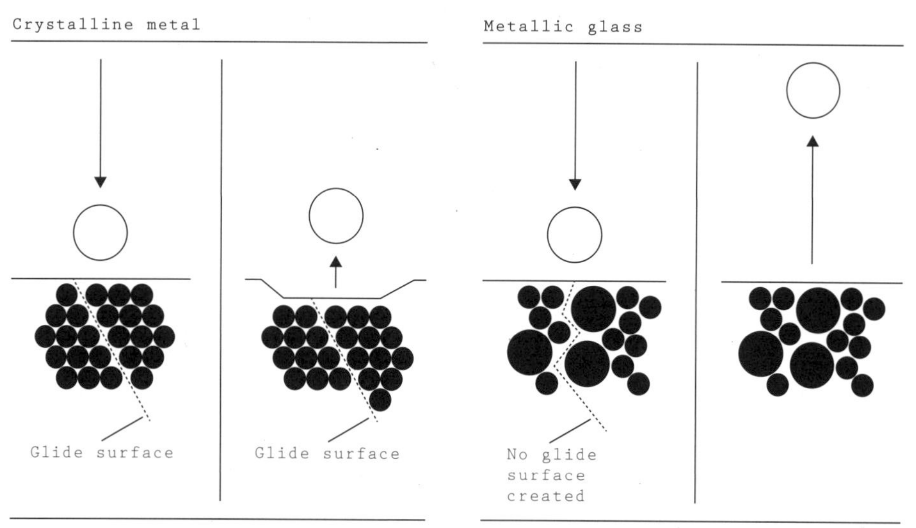

High-strength metallic glass for casings for electronic products (Source: Liquidmetal®)

At present, researchers are working on producing steels with an amorphous structure at significantly lower costs compared with the amorphous glass currently available. Being easy to process and highly corrosion resistant, as construction materials amorphous steels will compete with stainless steel and titanium.

Properties hydrophilic skin // moisture yielding // filtering of pollutants

Sustainability aspects efficient water production in dry regions

WATER-COLLECTING SURFACES

In November 2011, an extremely interesting irrigation system for the world's dry regions named "Airdrop" won the James Dyson Award. It is based on the principles of the Namib Desert beetle, whose hydrophilic skin could prove interesting for product developers and designers.

MATERIAL CONCEPT AND PROPERTIES

With its microscopic skin structure, the insect is able to obtain water from atmospheric humidity and thus survive even in dry desert regions. Droplets of dew stick to the skin, gather on the water-collecting surface, and drip along the insect's thick **chitin shell** and into its mouth.

APPLICATION

Australian designer Edward Linacre analyzed the principle of the beetle and transferred it to an irrigation system. Airdrop pumps air through an underground network of pipes, lowers the temperature to below the dew point, and extracts the moisture from the air for immediate use by plants. According to the developer's calculations, even in extremely dry regions the system could be used to extract up to 11.5 milliliters of water from one cubic meter of air.

At the ITV in Denkendorf, researchers are likewise working on technological applications for the functional principle of the Namib Desert beetle's skin. Here, in collaboration with the University of Tübingen, scientists are developing a **3D textile** that can be used to retrieve water from patches of mist in coastal areas. Depending on the region, one square meter of the new fabric can obtain from around three liters (Namib Desert) up to around 55 liters (South Africa) of water. Like Airdrop, the 20 mm thick material would be suitable for producing service and drinking water in dry regions. Moreover, the textile could be used to filter pollutants from the air.

Hydrophilic skin of the Namib Desert beetle (Photo: Thomas Schoch)

3D fabric for extracting atmospheric moisture (Source: ITV Denkendorf)

Airdrop irrigation system (Design: Edward Linacre)

Properties sound absorbing // based on biomass // outstanding sound properties in violins made of wood modified by fungi // sound can be influenced by liquids

Sustainability aspects lowering of acoustic stress // improved control of sound qualities

ACOUSTIC MATERIALS

Whether for optimum sound in a recording studio, the acoustics in a hotel lobby, or structural acoustics in an open-plan office, the correct use of acoustically effective materials is becoming increasingly important for interior designers and architects. The right materials have the potential to increase a sense of wellbeing, promote concentration, and reduce noise pollution at work.

Room acoustics are the easiest thing for architects or designers to change by varying their materials and material surfaces. Acoustic ceilings, textiles, or perforated wood-based materials can be used to influence the proportion of direct sound in the overall sound level, the time delay and direction of early reverberations, and both speed of onset and spatial range of echoes. The effect of materials on room acoustics depends on a number of factors and therefore each case must be considered separately. Lecture halls and theaters, for example, require particularly good speech intelligibility, whereas for recording studios the use of anechoic surfaces with low reverberation times is important.

For rooms in public buildings (kindergartens, schools) or offices, acoustic ceilings or suspended sound absorption elements are generally used to increase sound absorption. They absorb some of the incoming sound which in turn influences the reverberation time and room acoustics. The impact of acoustic materials is described as absorption coefficient α and has a value between 0 and 1. A value of 0.5 means that half of the sound is absorbed and half reflected. A value of α = 1 means that the sound reflection properties are completely cancelled out. The measures can be tailored to each situation in accordance with the size and arrangement of the acoustic material, the location of the sound source, and the direction of sound propagation.

The following acoustic materials are among the latest innovations in this specialist field:

NATcoustics — sound insulation with the biomass of the bulrush (Source: Naporo, photo: Diana Drewes)

Formation of spores of the fungi "Schizophyllum commune" in a petri dish (Source: Empa)

PRODUCTS

NAPORO NATcoustics

New regulations mean that action needs to be taken. Indeed, in Munich mineral wool may only be used to a limited extent as a sound absorber, and its unrestricted use in paneling in public buildings which children or young people frequent is prohibited. A bio-based alternative is NAPORO NATcoustics, based on the biomass of the cattail. The material can be simply plastered or covered with fabric and with a panel thickness of 50 mm has a weighted sound absorption coefficient alpha value of 1.0.

Banana fibers as acoustic material

In 2012, Maharani Dian Permanasari published the findings of studies on the use of banana leaves and bark fibers as acoustic materials. In a demonstration, banana leaves 20 mm thick reduced noises with frequencies of 200 hertz by 63%. In the high-frequency range, the absorption rate at 2,000 hertz was 55% and at 1,600 hertz 40%. The differences highlight the acoustic quality's clear dependency on the frequency of the soundscape.

Musical wood with biotechnological fungi treatment
Scientists at Empa are currently investigating how to achieve a significant improvement in the acoustic properties of wood used to make violins employing wood-rotting fungi. The ideal wood for a violinmaker has a low density, very good flexural rigidity, and enables high sound speeds during sound propagation. Investigating the fungi cultures, the scientists observed that the wood decayed slowly. In the later stages a wooden frame remains, enabling sound waves to spread better. The biotechnological treatment does not lower the flexural strength of the wood. The project, which is being supported until 2014, involves the production of 30 violins from fungi-modified wood with outstanding sound qualities.

Acoustic panels made of banana leaves (Source: Maharani Dian Permanasari)

Waterradio
By integrating thin metallic contacts in the surface of a piece of wood, designer Clemens Winkler has made it possible to prove the presence of spilt water. The water beads link the tiny contacts in the growth rings of the wood, which give rise to different frequency settings of a radio. As part of the installation, the listener is invited to play with the water and consciously manipulate the acoustic signals.

DILATANT FLUIDS

Dilatant fluids refer to substances whose flexibility and thus malleability change when force is applied.

Properties shock-absorbing // viscosity changes under application of force // intelligent modeling clay with magnetic properties // shock-absorbing foam // adaptive fibers

Sustainability aspects functional integration without additional components

MATERIAL CONCEPT AND PROPERTIES

The functional principle here is based on the atomic bonds in the **molecular structure** that form under the **application of force** and break again when the external pressure is removed. The **temporary linkage** increases the viscosity and lessens the fluidity of the substance. When strong pressure is applied, dilatant fluids abruptly become rigid. An example of a dilatant fluid that is easily made is **starch pulp,** a mixture of 400 grams of **starch powder** and 600 – 700 milliliters of water. Stirring the mixture reproduces the functionality of a dilatant fluid. When a whisk is moved gently in the pulp it seems liquid, but when the speed is increased it becomes brittle and thick.

Under normal conditions, dilatant fluids have viscous properties (Source: D30, Robert Palmer)

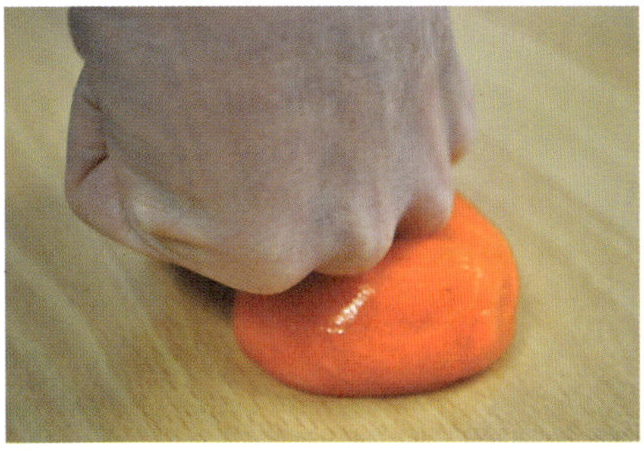

Dilatant fluids suddenly harden when force is applied (Source: D30)

APPLICATION

In the form of intelligent modeling clay, we find dilatant fluids both in toy stores and in physiotherapy practices. In recent years, dilatant fluids have been attracting increasing attention for technical applications. For instance, fibers immersed in them take on shock-absorbing and particularly durable properties. The phenomenon is currently being investigated, particularly in combination with carbon fibers for shock-absorbing textiles. Moreover, dilatant fluids have potential for use as **intelligent lubricants** in mechanical engineering.

SPECIAL DILATANT FLUIDS

Intelligent modeling clay

As early as the 1940s, US researchers looking for an alternative for **natural latex** discovered this **intelligent modeling clay based on silicone,** whose degree of malleability changes under the application of force. It has been extremely popular with children ever since, as it can be worked by hand into all imaginable shapes. Yet when thrown on the floor, the modeling material bounces back very quickly. Under great pressure it becomes brittle and breaks up. Alongside its use as a toy, intelligent modeling clay is employed in model making and ergotherapy. It doesn't dry out even after hours of use. Intelligent modeling clay is sold in colors that change tone with temperature or with magnetic qualities.

Shock-absorbing polymer foam D30

British manufacturer D30 specializes in the manufacture of an elastomeric polymer using a shock-absorbing dilatant fluid. Under normal conditions, the material displays sufficient flexibility which, however, it abruptly loses when subjected to a strong external force. The viscosity of the fluid changes, the material hardens and thus absorbs shocks and deformations. When the force is removed, the material regains its flexible properties. D30 is suitable for use in anything from sports clothing and protectors to helmets, bulletproof vests, and shockproof electronics housing. The first applications can primarily be found in the ski and winter sports sector. The three D30 qualities ST, XT and Shock, which differ in terms of hardness and tear resistance, are available on the market.

Dilatant fibers

In the context of the project **"Rheocore,"** Swiss researchers at the Empa are currently developing **adaptive plastic fibers** using a dilatant fluid in a duct system integrated into the fibers. The fibers are suitable both for adaptive applications and for ultralight composites.

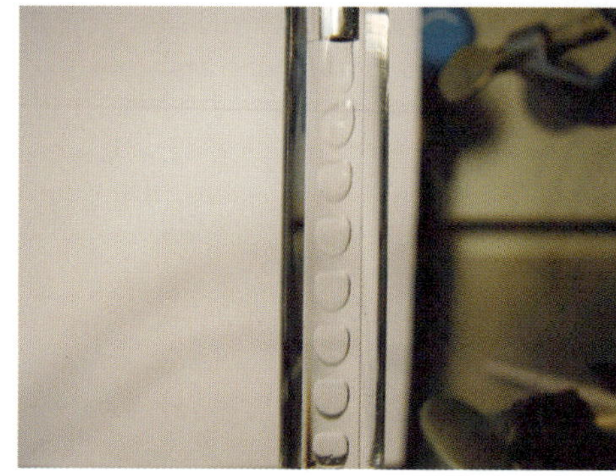

Active fibers being filled with a dilatant fluid (Source: Empa)

Structure of a shock-absorbing textile using D30 (Source: D30)

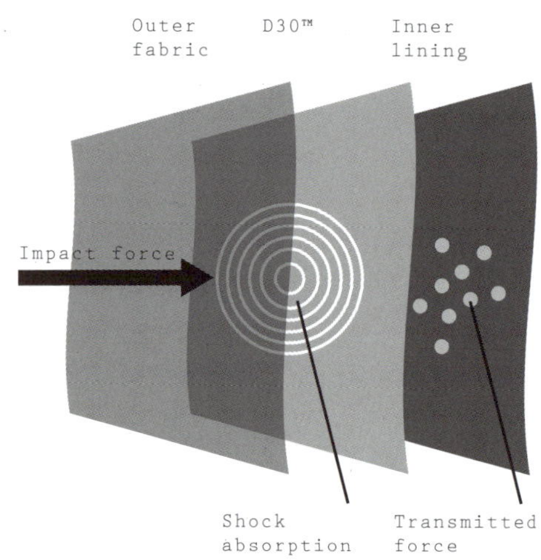

Properties changes form in an electric field // tactile information // vibration damping // energy harvesting from vibrations

Sustainability aspects able to generate energy

ELECTROACTIVE ELASTOMERS

Today electroactive plastics in the form of foils that change shape when a voltage is applied are no longer something just used in research, but are in practical application. One example is an electroactive composite foil for cell phone displays, which changes shape when electrical voltage is applied and gives the user a tactile control signal.

Since 2010, engineers at the Fraunhofer LBF in Darmstadt have been developing an electroactive elastomer that either absorbs unwanted vibrations or is able to supply sensors in inaccessible places with electricity.

MATERIAL CONCEPT AND PROPERTIES

To present the functionality of the principle, the researchers have developed a demonstrator composed of 40 thin **elastomer electrode layers** that is smaller than a cigarette packet. It is even possible to produce electricity by reversing the function of the **stack actuator**. This property is interesting, for instance, in terms of monitoring inaccessible areas where there are vibrations, but no power connection.

APPLICATION

Integrated into a bicycle saddle, the new material could, for instance, alleviate cycling on cobblestones. Moreover, the engineers at the Fraunhofer Institute see a possible use for the stack actuator in vehicle construction. Although engines are positioned carefully, the use of **active elastomers** can reduce vibrations in a car even further. Researchers at the Hamburg University of Technology are working on active elastomers for the padding of automobile seats that can be adapted to the body shape and size of the user.

Structure of a stack actuator consisting of electroactive elastomers (Source: Fraunhofer LBF)

Functionality of an electroactive display for cell phones (Source: Bayer MaterialScience)

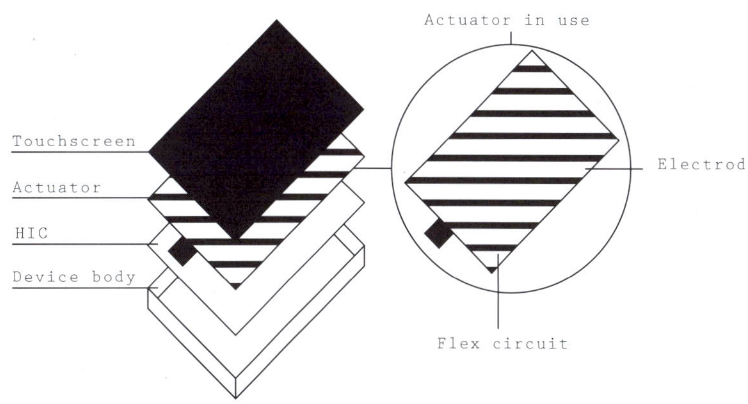

Materials and surfaces that can change their shape or geometry when an external force is applied are highly interesting as fields of development among functional materials. Shape memory materials and bimetals, for example, respond to heat and are suitable for various uses in the control of devices and interior design. Whereas electroactive plastics contract or expand when a voltage is applied, hydrogels react to moisture. Expancel microspheres are also giving rise to a whole new form of aesthetic changeability.

Properties changes volume when heat is applied // sound-insulating effect // extremely light additive // increases durability

Sustainability aspects potential as lightweight construction material // heat-insulating qualities

EXPANCEL MICROSPHERES

MATERIAL CONCEPT AND PROPERTIES

Expancel microspheres are extremely small **hollow spheres** that can be mixed with surface coatings or plastics as an additive. When heat is applied, the **internal gas pressure** of the spherical particles increases, leading them to increase considerably in volume. The polymer shell softens and expands to over 40 times the original volume. This effect can be used, for instance, to create thermovariable **texture changes** on printed layers or plastic packaging. Integrated into a matrix, the **hollow spheres** are also heat and sound insulating.

APPLICATION

Expancel microspheres can be used in anything from the manufacture of shoe soles in plastic injection molding, to the production of tennis balls, artificial marble, or bottle corks. As an extremely light filler material they can also be used in car bodies and ship hulls. Expancel microspheres are available in both expanded and nonexpanded form. Their industrial applications include as blowing agents to increase durability with simultaneous weight reduction, to improve ease of application, and for surface finishes.

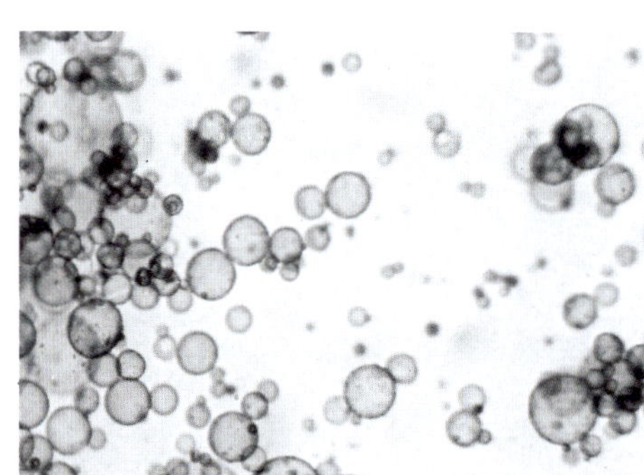

Volumechanging microspheres (Source: AkzoNobel)

AUXETIC MATERIALS

Properties negative Poisson's ratio // high energy absorption // good breaking strength // release of active agents in auxetic composites

Sustainability aspects high material efficiency // lasting structures

So-called auxetic materials can be deployed to increase the safety of certain applications in sports and industrial production. This is a group of materials that has hitherto received little attention from researchers.

MATERIAL CONCEPT AND PROPERTIES

Auxetic materials are materials, fibers, or foams that get thinner under pressure and thicker when stretched. They have a **negative Poisson's ratio,** meaning that they behave in exactly the opposite way to most other materials. The phenomenon can be seen both at the molecular and macro level. One example found in nature is **cow teats.**

APPLICATION

Auxetic materials have increasingly become the focus of research activities recently and the potential applications are manifold. The materials can absorb a great deal of energy and are particularly resistant to breakage, making them potential options for bulletproof vests or sound and shock absorbers. In addition to safety equipment, in medicine **auxetic composites** could release active agents when a wound becomes swollen. In the construction sector, auxetic materials have potential for use in earthquake-proof construction. They would also be suitable for use as bone substitutes, implants, or artificial lungs.

Principle of auxetic materials
(Source: University of Malta)

Normal materials

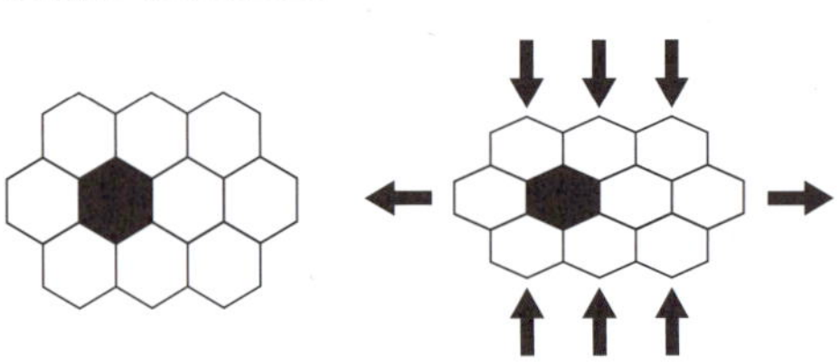

Auxetic materials

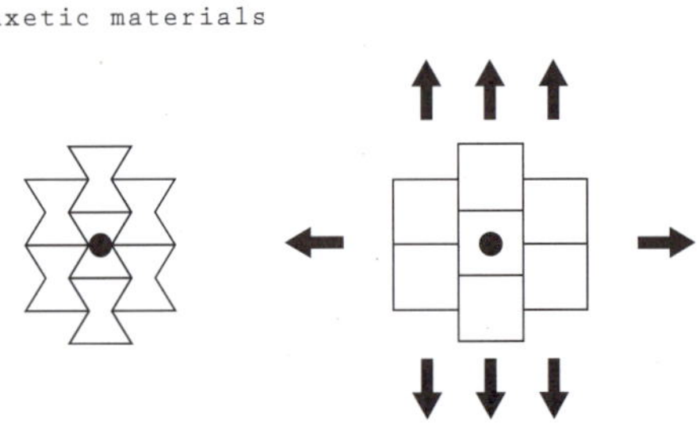

Lattice structure with auxetic qualities
(Source: University of Erlangen-Nuremberg)

Fiber strand with auxetic properties
(Source: Danish Design Centre)

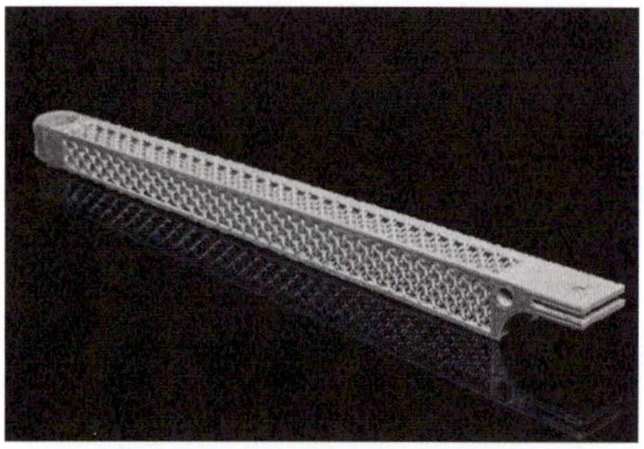

Robot arm with auxetic structure
(Source: University of Erlangen-Nuremberg)

Materials with shape memory properties have been around for some time as nickel or titanium-based metal alloys, for instance for artificial heart valves. They are materials that return to their original shape when heated to a certain temperature. On the back of a new development by Bayer MaterialScience and the Federal Institute for Materials Research and Testing (BAM), it is now possible to produce technical components made of TPU with memory effect.

Properties temporary shape-setting following targeted thermomechanical treatment // reversible when heated to over 40 °C // thermoplastic processing possible

Sustainability aspects integration of an additional function with no extra material // improves ease of assembly // no softening agents

THERMOPLASTIC POLYURETHANE (TPU) WITH SHAPE MEMORY

MATERIAL CONCEPT AND PROPERTIES

Shape memory TPU can be shaped using injection molding or extrusion. These methods enable a whole range of industrial products to be made that, when subjected to a targeted thermomechanical treatment, can be temporarily set in a different form. The component regains its original shape when heated to a temperature of 40 °C. The shape memory effect is not specific to TPU, but is the result of a particular molecular structure and can also be used for dual programming.

APPLICATION

Possible applications include in the fields of trademark protection and mechanical engineering as well as in the textiles, sports, leisure, and toy industries. Products that could be made from shape memory TPU include artificial muscles, temperature sensors, hinges, shrink tubing, self-loosening screws, and textiles that "de-crinkle" themselves. One patent-pending product idea is a greenhouse-like **foil tunnel** for vegetables and salad cultivation that moves into the vertical under exposure to heat. Research is also under way on **forgery-proof labels** with a QR code. The new plastic is suitable for contact with food, has no softening agents, and is resistant to chemicals.

Forgery-proof labels with QR code (Source: BAM, Thorsten Pretsch)

Functional principle of a foil tunnel made of shape memory TPU for vegetable and salad cultivation (Source: Bayer MaterialScience)

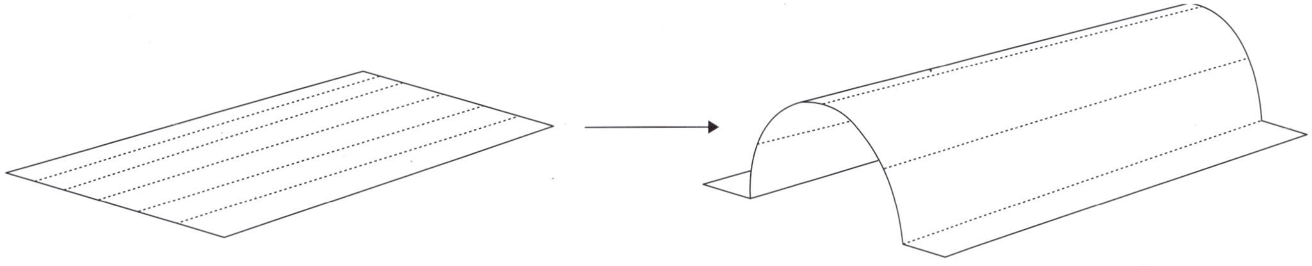

NANOPOROUS GOLD

Properties moldable under normal conditions // rigidifies under the influence of an electric field // mechanical strength doubles

Scientists at the Hamburg University of Technology and Chinese Academy of Sciences have developed a material called nanoporous gold which has shape-changing and mechanical properties that can be controlled by applying an electrical voltage.

MATERIAL CONCEPT AND PROPERTIES

Nanoporous gold consists of a sponge structure with openings between 10 and 20 nanometers that are filled with **perchloric acid.** The pores and ducts are seared into the material and enlarge the surface in relation to its volume. Under normal conditions the material is flexible and can be molded. When a voltage of 1.5 volts is applied, however, it becomes rigid and set. The reaction is down to a layer of **hydroxide ions** that accumulates on the **surface of the gold** and doubles its mechanical strength. The ions come from the acid and detach from the surface again as soon as the voltage is switched off. The **gold oxide layer** disappears and the material regains its original plasticity.

APPLICATION

Scientists are still researching the cause behind the phenomenon. They assume that the layer of ions prevents irregular movement in the crystal lattice and causes the increase in rigidity. Initial applications for the intelligent material are already under discussion. The effect has also been observed in platinum.

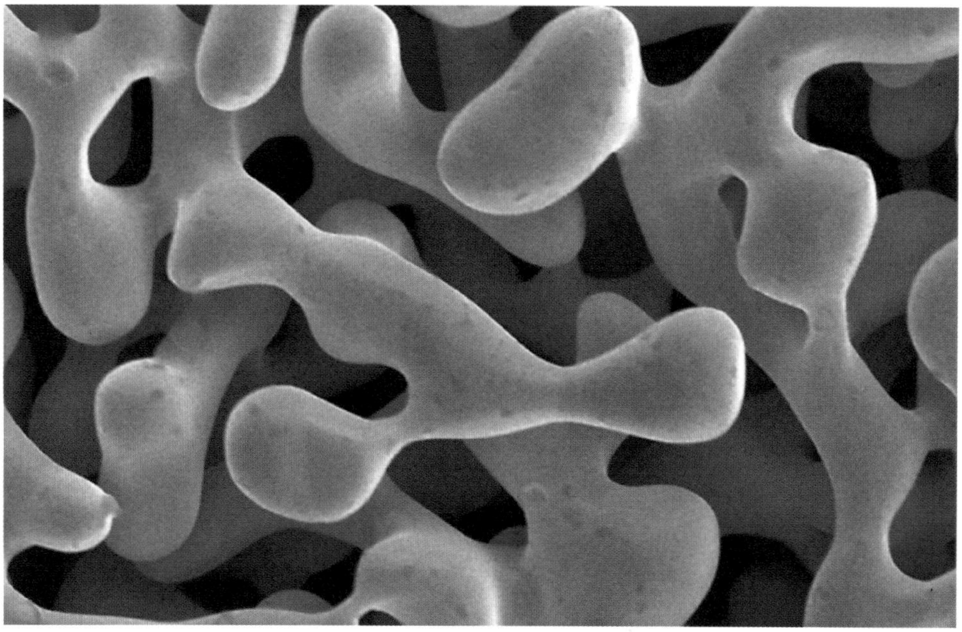

Microscopic image of nanoporous gold (Source: Helmholtz-Zentrum Geesthacht)

GRADIENT MATERIALS

Properties continual adjustment of properties // exposed concrete without insulation possible // fluid metal structure transitions // textiles with smooth coating

Sustainability aspects increase thermal insulation properties // material-efficient solutions // avoidance of excess material

Gradient materials are those whose properties can be adjusted accurately and continuously, and tailored to their particular use. Most notably, solutions have been developed as prototypes for the construction industry. Numerous uses are also expected in the fields of aviation and energy technologies. In addition to their focus on the properties of various materials, researchers are also investigating sealing layers and techniques for heterogeneous composites (for example, metal/glass or concrete/wood).

Gradient concrete

Researchers at the University of Stuttgart are currently developing gradient concrete as a new class of building material whose porosity can be continuously adjusted. By specifically influencing the size of internal hollow spaces, its characteristics, i.e. strength, thermal insulation, and density, can be modified with great precision. In this way gradient concrete could be used optimally in relation to resource consumption and be precisely adjusted to the particular requirements in each case. The aim is to avoid excess material. According to calculations by scientists at the Institute for Lightweight Structures and Conceptual Design (ILEK) at the University of Stuttgart, exact alignment with the load curve would reduce the material weight of ceiling slabs by up to 60%. Insulating layers in the core could influence the thermal conductivity such that exposed concrete could be used without any additional insulation. Work is also under way to improve the concrete properties in sandwich constructions with membranes and foils with the integration of **PCM** or **aerogel.**

Gradient metals

In the development of gradient metals, scientists are primarily concentrating on the functional adjustment of the qualities, which they can achieve in particular by means of the alloy composition or by influencing the material structure. They aim to create highly stressed components for the aerospace sector, for energy systems, or to achieve certain thermal conductivity properties. In particular, a fluid phase transition eliminates weak points stemming from different material qualities at the interfaces. Methods from **powder metallurgy** have proven their value for production. One option is the **sedimentation process,** used to transform metallic and nonmetallic powders into so-called **gradient green compacts** with fluid **structural transitions.** Conventional sinter technology is then employed to achieve the required densification. Given the increasing significance of generative production methods for direct product manufacture, in **laser sintering** the powder mixture could be precisely matched to the relevant application. In theory, this technique would also work with plastics or ceramics.

Gradient plastics

With a view to the construction industry, researchers have already produced structures from open-cell foam with a fluid porosity curve. In order to subsequently lend them sufficient strength, they infiltrated them with other materials and bonding agents.

Gradient textiles

For the textiles sector, solutions have already been found with a smooth coating that augments rigidity, adjusts the **permeation properties,** and step-grades the **moisture transfer** in various areas. This is already state of the art in the textile industry, while in the construction sector there are different definitions of moisture transfer in the building shell. Moreover, differences in rigidity could be continuously and seamlessly defined and achieved.

Gradient concrete
(Source: University of Stuttgart, ILEK)

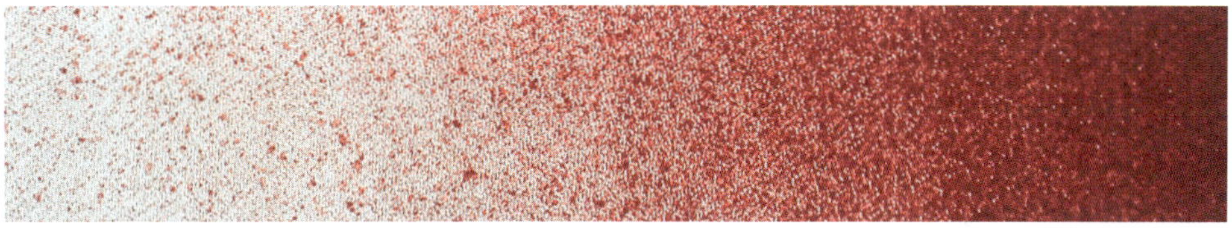

Fiberglass fabric with continually increasing concentration of silicone, for staggered permeability level
(Source: University of Stuttgart, ILEK)

Smooth transition between concrete and wood
(photomontage)

Properties continual adjustment of properties // exposed concrete without insulation possible // fluid metal structure transitions // textiles with smooth coating

Sustainability aspects increase thermal insulation properties // material-efficient solutions // avoidance of excess material

METAMATERIALS

Metamaterials are materials with unusual properties that are not encountered in this form in nature. The arguably best-known effect is based on the refraction of light in a particular way, making objects appear invisible. In recent years research activities in the field of metamaterials have intensified substantially, and there are now various points of focus.

PRODUCTS

Electromagnetic metamaterials

Most research projects focus on working with artificial materials that have a **negative refractive index** and direct waves around an object. To this end, structures are generated in the surface of the material that are smaller than the wavelength of visible light. Research on the nanometer scale often requires new production approaches. For instance, a method is currently being developed at the Karlsruhe Institute of Technology (KIT) called **Direct Laser Writing** (DLW), which is intended to enable the generative production of tiny **3D structures**. This method has been used to realize the first **3D invisibility cloak** for unpolarized visible light in the range of 700 nanometers.

Using a **plasmonic metamaterial,** in January 2012 US researchers in Texas were also able to make a three-dimensional shape appear invisible. The scientists used a metamaterial that scatters light rays in exactly the opposite way to conventional material surfaces. The waves overlap and cancel each other out. To date, however, the invisibility cloak only works for an 180 mm cylindrical tube in the microwave range.

Super lenses

Super lenses are optical lenses that have no **resolution limits.** According to **Abbe's diffraction limit,** the best optical resolution is limited to around half the wavelength of the light used. By exploiting nanostructures, scientists could change a lens's reflective behavior to create significantly smaller and far more sensitive optical sensors and lenses for the fields of medicine and environmental technology. Super lenses would also not be curved, but have a flat shape.

Scientists at the Technical University Munich (TUM) and Ludwig Maximilian University of Munich (LMU) are working on a super lens consisting of DNA molecules. The molecules automatically arrange themselves along a spiral and are combined with **gold nanoparticles** enabling them to influence light rays. The metafluid has been successfully used to completely filter green light irrespective of the direction of the radiation. As such, the scientists can now create nano-optical materials with precisely specifiable qualities. In the long term, the metafluid has the potential to enable the creation of a "perfect lens" and is also relevant to the realization of an invisibility cloak.

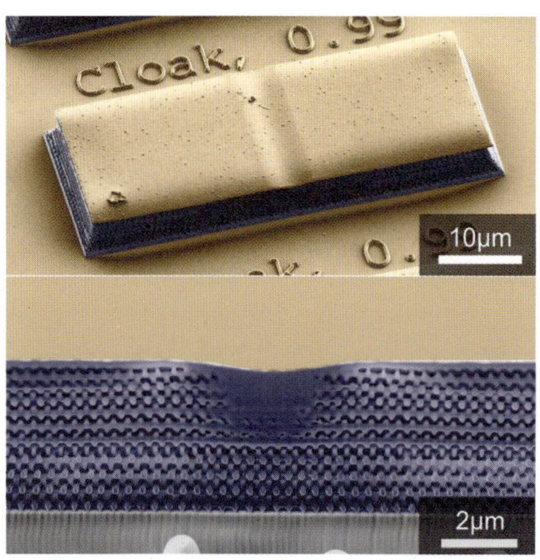

Electron microscope image of an invisibility cloak structure for 3D structures (Source: KIT)

Structures produced by means of Direct Laser Writing (Source: KIT)

Seismic metamaterials

Seismic metamaterials are being developed to reduce the vibrations triggered by earthquakes. The idea is to divert **surface waves,** which are responsible for the greatest damage, around buildings by means of a bulwark of rings anchored in the ground. The seismic **invisibility cloak** currently being developed at Fresnel Institute in Marseille, for instance, consists of at least 10 rings which together form a metamaterial. When a seismic surface wave hits the first ring, it triggers a certain interaction between the ring and the wave. The ring bends and performs a countermovement, which diverts the wave. The individual rings are set up for different frequencies of surface wave, meaning they can influence a wide range of waves.

Acoustic metamaterials

The University of Illinois has come up with a solution for an acoustic metamaterial that could also be used for objects in larger dimensions. A metallic ring with a diameter of 100 mm and 16 **concentric ring structures** can be used to block the perception of waves in the ultrasonic range between 40 and 80 kilohertz (e.g. sonar waves). The speed of the sound waves is changed by means of the geometry of the rings. They spread out in the **hollow chamber structure,** are slowed down, and absorbed. Researchers see initial applications for the material in the military and medical fields. For example, the ring structure could be put to good use on submarines in rough seas. It could also be used to protect oil rigs from the forces of nature. And it could eliminate noise pollution from concert halls or vehicles.

Metafluids

With their stable crystalline **metafluid,** a **pentamode metamaterial,** scientists at the KIT have successfully developed a new class of materials. They are produced by means of new **nanostructuring** methods and can take on every conceivable mechanical property. As regards parameters, the ideal state of a pentamode metamaterial corresponds to the state of water. Water can hardly be compressed in a cylinder, but can be stirred with an object. The mechanical behavior of the metafluid is determined by how pointed and long the individual hat-shaped elements in the diamond-like molecular structure are. The material mass only occupies around 1% of the body volume.

Organic metamaterials

In late 2012, US researchers at Cornell University announced the discovery of a new metamaterial phenomenon. Using organic substances, the scientists succeeded in producing a material that is liquid when dry and assumes solid form in a liquid. The material was produced from **deoxyribonucleic acids** (DNA) in the three-dimensional network structure of a **hydrogel** with excellent water-retention properties. One potential application for the organic material is water-reactive electric switches.

An aluminum disk structured into a metamaterial can refract ultrasonic waves around the free central area (Photo: L. Brian Stauffer, University of Illinois)

Dichroism of DNA lenses two beads with the optically active DNA gold molecules absorb polarized light to different degrees (Source: TUM/LMU)

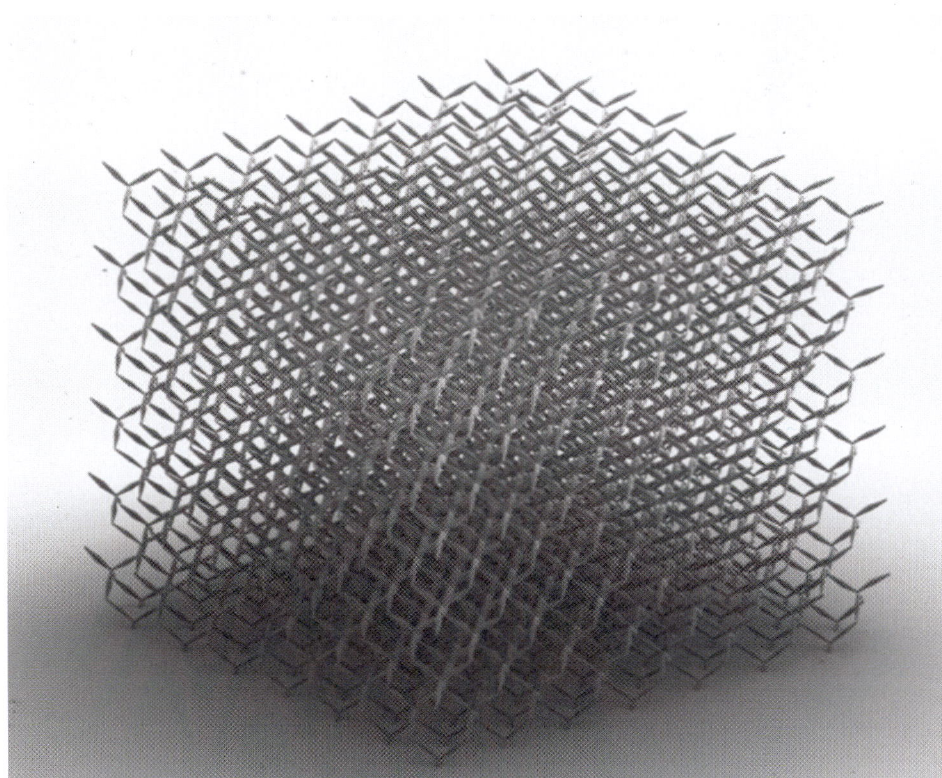

Pentamode metamaterials behave in a similar fashion to liquids (Source: CFN, KIT)

MATERIALS THAT INFLUENCE AND EMIT LIGHT

06

Over the last few decades, artificial light has shaped the spaces we live in more than virtually any other medium. Within this process, the dividing lines between lighting and sources of information have become ever more blurred. In place of light bulbs we now have tiny LEDs (light-emitting diodes) or ultraflat OLEDs (organic light-emitting diodes). Nanotechnology has firmly established itself in lighting technology. Semiconducting diodes and illuminating light surfaces now need very little material. The light-emitting technology merges with the architecture. At the same time, material solutions with transparency-changing, light-directing, and variably reflecting qualities are shaping our perception of both interior and external light. New possibilities for interaction link the diffusion of light rays with forms of human movement.

Increasing energy efficiency and ease of production are the main focus of most development processes in lighting. Developers discern immense potential here by optimizing printing technology for OLED dyestuffs. In future, the creation of textile-based organic light sources will also open up new creative possibilities for designers, such as the enhancing of conventional reel-fed printing processes to incorporate light-emitting electrochemical cells. Recently, designers and architects working on energy efficiency have transferred examples to the sphere of artificial light in a series of projects using bioluminescence phenomena. Along with the particularly aesthetic quality of their natural light, photoproteins and bacteria show a dramatically high level of efficiency as regards the conversion of energy into light.

MATERIALS THAT INFLUENCE AND EMIT LIGHT

Optical Textiles

156

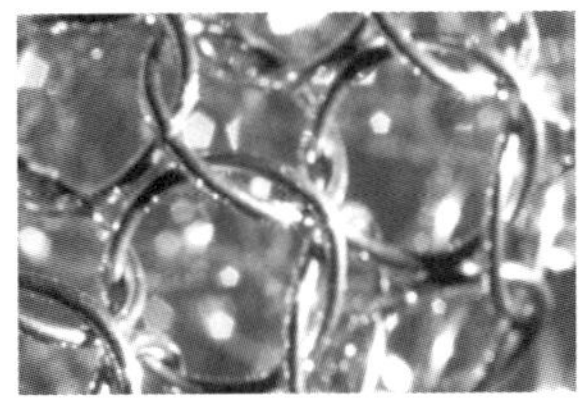

Light-reflecting Metal Ring and Metal Flake Meshes

160

Interactive Light

164

Polymer Optical Fibers (POF)

157

Antireflective Coatings

161

Light-emitting Electrochemical Cells (LEC)

166

Transparency-changing Materials

157

Led Media Materials

162

Biological Light

166

Light-directing Materials

159

Electroluminescent Materials

163

OPTICAL TEXTILES

The significance of technical textiles in architecture has consistently grown over the last few years. On the one hand, this is down to the material's low weight and flexibility, but these textiles can also be used to great effect with regard to light. A particular challenge when it comes to working with optical textiles is the material's even light distribution.

Properties uniform back-lighting with various light sources // low structural height // PTFE fabric with good light transmission for eternal use // acoustic qualities

Sustainability aspects light distribution with no extra technical effort // low weight // quick to mount

PRODUCTS

Translucent wall and ceiling linings

As part of a development project, PONGS® Technical Textiles and li`ccon® created a textile solution for wall and ceiling linings with a translucency of 35%, which provides **uniform background lighting** with **LEDs** and shields, offering sideways irradiation without the position of the light sources needing to be visible. The maximum structure height is 800 mm.

Barrisol® light-diffusing stretch ceilings

The flexible **stretch ceiling system** consists of a profile system and polymer fabric. The thermoplastic fiber material becomes more malleable when heated and is stretched over the frame construction. As it cools, the fabric tightens slightly and the stretch ceiling takes on its final form. If the material is lit from behind, then the special texture of the fabric means that the light is distributed particularly uniformly, creating the impression of a **flat light source.** There is a choice of 13 different colors and transparency levels, so the stretch ceilings can be used for virtually any ambiance required.

Light-engineered polytetrafluoroethylene (PTFE) fabric

The special fabric manufacturer Sefar offers PTFE fabric for lighting technology installations both internally and externally. For interiors, these boast a very high degree of **scattered light** with minimal color shift (light transmission: 80 – 85%) and also have a sound-absorbing effect. PTFE fabric with good tensile strength offers sufficient weather resistance for use outdoors. For **facade decoration** in particular, lightweight, open-mesh fabrics with superior aesthetic properties are available.

Rice Fold

This folding **acoustic ceiling** is based on recycled polycarbonate made of old water containers and CDs. It is light-permeable and optimized for use with LED light. The strong light distribution makes the material suitable for covering ceilings and walls. The geometry is designed so that Rice Fold can be installed in the shortest possible time. A ceiling construction requires neither screws nor adhesive and one unit weighs in at just 1,100 grams.

250 highly flexible sunshades made of 143,000 m² PTFE provide plentiful shade at the Medina Haram Piazza (Source: Sefar)

Folding structure of the Rice Fold acoustic ceiling (Source: Miniwiz)

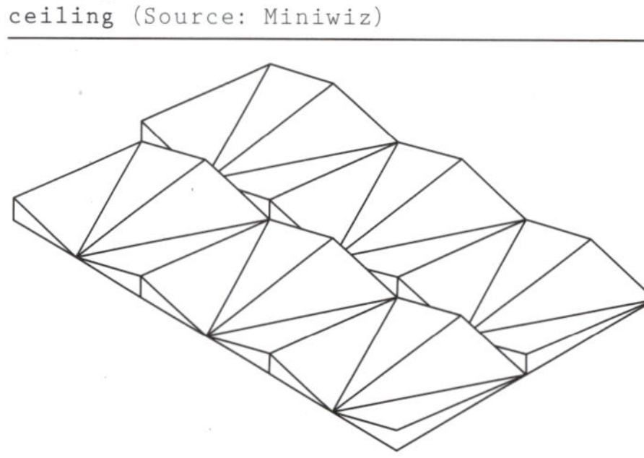

POLYMER OPTICAL FIBERS (POF)

Light-conducting fibers are something we are particularly familiar with in the form of glass optic fibers, which can conduct light from one end to another. However, there are also light-wave conductors made of synthetic materials, which can be used straightforwardly in furniture construction, accessories, and game design.

Properties light-wave conductor based on optical synthetics // low weight // high flexibility // highly transparent fiber core coated with a low light-refraction material // channeling of the light rays to the boundary layers through reflection

Sustainability aspects light conducting with no great effort // low weight

MATERIAL CONCEPT AND PROPERTIES

These come under the umbrella term **"polymer optical fiber"** (**POF**). They generally consist of polymethyl methacrylate (PMMA), an optical plastic most commonly known under the brand name "Plexiglas." The highly transparent **fiber core** is coated with a very thin layer of **fluoride PMMA** with low light refraction. Light rays are channeled to the boundary layers through reflection and are even conducted around curves without loss. The low **refraction index** in the coating prevents light from being radiated. Along with their low weight, polymer optical fibers also boast a high level of flexibility. Compared with glass fibers, though, their usage temperature is limited to 60 °C.

APPLICATION

Optical cables are extremely important for data transfer. When it comes to fashion, POFs create impressive effects. Light appears everywhere where the fluoride PMMA coating is interrupted. Interior designers use optical fibers, for example, in textile drapes or as room dividers. Under the name "Mist Bench," designer Gwenael Nicolas used POF to create a seat that reacts to the movement of its user. It consists of optical fibers that have been knotted by hand to create a sufficiently robust surface textile. By means of a sensor, specific areas of the bench are triggered so that the light intensity is increased when a person approaches. In dim light this gives the impression of a misty shroud.

Mist Bench
(Design: Gwenael Nicolas)

TRANSPARENCY-CHANGING MATERIALS

Properties change in transparency resulting from the influence of heat or electric current // transparent properties possible for leather and wood // opaque glass at the touch of a button

Sustainability aspects change in transparency by simple means // use of low-voltage current // self-regulating glare and heat protection

Materials, films, and surface coatings that change their transparency as a result of external influences have gained considerably in significance over the last few years in intelligent material applications in design and architecture.

PRODUCTS

Thermosensitive fat

A particularly impressive example, realized with very simple means, comes from designers in the Netherlands. The Slow Glow Lamp consists of a spherical glass bulb, in the middle of which is a light bulb surrounded by **animal fat.** When the lamp is turned on, the resulting heat melts the fat, which then takes on a light-permeable transparency, making the light bulb clearly visible. Once the lamp is turned off, the fat gradually cools down and regains its opaque consistency.

Liquid crystal foils

A particularly widespread process for the careful control of transparency in glass is the use of **liquid crystal foil.** It is applied between the two sheets of a **laminated pane of glass** and reacts to the application of an **electric current** with a specific alignment of crystals that makes it transparent. As soon as the current is interrupted, the orientation of the crystal structure is lost. Rays of light are scattered widely across the **liquid crystals** and the glass appears opaque. Liquid crystals are immensely important for self-build displays and advertisements; as a consequence the relevant instructions are readily available on the Internet.

Ynvisible™

As an alternative to liquid crystals, economical solutions for interactive displays and advertisements that make use of electrochromatic print colors and conventional print technology are now available on the market. Possible applications of the technology therefore include not only interactive games and greeting cards, but also **intelligent packaging, e-paper, interactive clothing,** and smart furniture surfaces.

SageGlass®

Switching on an electric impulse can change the transparency properties of this glass for adjustments to windows, skylights, and rainscreen cladding or to provide protection from glare in the summer without the need for blinds. The efffect is based on five wafer-thin **ceramic layers** with a total thickness of less than a 50th of a human hair, which are placed between two sheets of glass. Applying a **low-voltage current** results in a reorientation of electrons and ions and thus a change in transparency. Light and heat rays are absorbed and then re-emitted from the glass surface. SageGlass® is available on the market as **two** and **three-layer insulating glass.**

Glass with photochrome resins

In cooperation with Tilse Formglas, Fraunhofer IAP has developed a glass solution with transparency that changes under the influence of light and heat. To create this effect, scientists placed a **resin layer** containing **microcapsules** from a synthetic material between two glass sheets. When the capsules are heated to 40 °C they change their structure and falling rays of light are diffused. In this way, 30 – 50% of the sun's energy can be radiated back outwards again.

Transparency-changing wood and leather

A process developed by industrial designer and model-builder Hauke Reiser makes an opaque material like leather or wood transparent with reverse lighting, so that graphics or animations underneath the material become visible on its surface. This process makes it possible to create exciting light effects, which are unexpected in their form given the opaque nature of the materials. The technique can also be used on three-dimensional surfaces and so, for example, in the automobile and furniture industries.

Layered structure of the patented interactive display (Source: Ynvisible™)

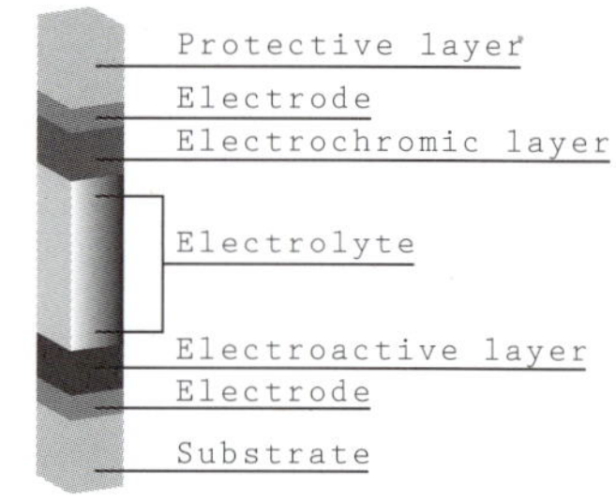

Back-lit transparent wood (Source: Hauke Reiser)

SageGlass® with electrochromatic properties (Source: Saint-Gobain)

Slow Glow Lamp with thermosensitive animal fat (Design: Next Architects and Aura Luz Melis)

Over the last few years, certain materials and material surfaces have been developed for interior applications that conduct light through solid walls, focus rays on one area, or break up the light with varying levels of intensity to reveal the different colors of the spectrum.

Properties light-permeable concrete // color effects with dichroic glass // light direction with nanomirrors // bundling of light rays with Fresnel lenses // increase in energy yields from PV systems

Sustainability aspects increase in yield from PV systems // climate-optimized light control

LIGHT-DIRECTING MATERIALS

PRODUCTS

LiCrete®

This new construction material consists of 85% concrete and 15% transparent materials. Its particular translucence makes it suitable for setting up back-lit walls, staircases, or counters made of concrete. At the same time, production of LiCrete® is much more economical than the comparable offering of **light-permeable concrete.** Given the favorable ratio of concrete to the lighter additive, the material boasts considerable stability. Its mechanical properties mean it can be used by architects and designers as a construction material. Products are also available that allow light to enter from the front or back and exit from the side surfaces.

Holographic optical elements (HOE)

With an **HOE,** light is broken up at various levels of intensity; depending on the angle at which it falls, it becomes visible in its various spectrum colors according to the wavelength. For some wavelengths, at a specific angle the elements are entirely transparent, while for others they are not. The overlapping of several **holograms** means rays of light can be diffused in different directions. When embedded in glass facades, **HOEs** achieve impressive effects in architecture and design. Technical applications of HOEs include scanners, displays, and credit cards, for example. Compared with classic glass lenses, HOEs have extremely low thickness, since they are manufactured through illumination of a **holographic film** with a **laser beam.**

Dichroic glass

Dichroic glass refers to color-effect types of glass that break up the light to varying extents and make the different colors of the light spectrum

Light-permeable and light-directing concrete building blocks (Source: Gravelli)

Screen "And A And Be And Not" made of dichroic glass (Design: Camilla Richter)

visible depending on the angle and the type of light source. The filters are applied to the glass as **metal oxide layers** with coating solutions in an immersion procedure and burnt on at a temperature of 480 °C. Dichroic glass is highly resilient and is therefore suitable for architectural applications as well as for interior design and furniture construction. One example is a folding screen made of types of dichroic glass, which was presented at imm cologne in 2013.

Fresnel lenses

Originally invented and developed for lighthouses at the beginning of the 19th century, **Fresnel** lenses concentrate and bundle rays of light. In the field of optics, they are also used in projectors, **solar-thermal systems,** or to increase the energy yield of **photovoltaic systems.** Thus the use of light-concentrating lenses in the laboratory has already resulted in efficiency factors of 40% being measured on PV systems. This is twice the value of the normal yield.

Liquid lenses

The combination of cylindrical lenses in crossed alignment with water and the resulting bundling of the sun's radiation on the photovoltaic systems permit a further increase in the efficiency factor. Lumicell CPV-Systems from Berlin is the world's first supplier of liquid lenses.

Nanomirror for smart windows

At the University of Kassel, scientists have experimented with the integration of **nanomirrors** (so-called **micromirror arrays**) in the space between sheets for the development of intelligent windows for controlled **light-direction** of daylight to specific places within the space. This function is to be achieved with around 10 million minute, but electrically movable mirrors on a surface of one square meter. The scientists are also looking at using the nanomirror for color projections in order to project images onto the wall. Every mirror would then correspond with an image pixel.

Nanomirror for climate-optimized light direction (Source: University of Kassel)

Fresnel lenses made of PMMA for concentrating PV (Source: Concentrator Optics)

Liquid lenses increase the efficiency factor of a photovoltaic system (Source: Lumicell CPV-Systems)

Properties reflection effects on metal rings and flake weave fabrics // light-permeable // interior and external applications

Sustainability aspects light effects with no additional electronic input

For facade construction, metal surfaces not only offer protection against outside influences, they can also create particular light effects and ambiences by means of meshes and ring structures.

LIGHT-REFLECTING METAL RING AND METAL FLAKE MESHES

MATERIAL CONCEPT AND PROPERTIES

Facade and surface materials consisting of **metal ring** or **metal flake meshes** are available on the market under the brand "Alphamesh." The material is made of thousands of rings woven together, and is as suited to use as a **facade element** as it is for decorative use in interior design. As soon as the light hits the highly reflective metal surface, atmospheric scenes are created.

While **metal ring fabric** is supplied in **stainless steel** or **bronze** with standard ring diameters of 7 mm and 12 mm, metal flake fabrics in brass and aluminum are also available. The basic weight of a stainless-steel woven fabric with a ring diameter of 12 mm is just 3.2 kg. Flake fabrics are available in a maximum format of 3×6 m and are suitable for external use. A total of 27,750 flakes are needed for the production of every square meter.

APPLICATION

The transparency of the ring structure means it is possible to work with light effects and create scenic atmospheres in architecture and design. The combination of spatial flexibility, light conduction, and playing with transparency offers an alternative to the cumbersome nature of conventional metal surfaces. The variety in production means that the company can produce structures tailored to individual requirements.

Metal ring structure (Source: Alphamesh)

250 m long metal ring curtain at Swarovski in Wattens (Source: Alphamesh, Photo: Daniel Swarovski)

ANTIREFLECTIVE COATINGS

With both solar panels and PV elements, sunlight is reflected to a not insignificant extent, resulting in a reduction in the efficiency of the system. Assistance is provided in the form of antireflective layers, which are now incorporated into a good 10% of all solar panels sold. With the aim of increasing effectiveness, companies and scientists are continuing to work on ways to further reduce reflective properties.

Properties reduction in reflective properties // structured surfaces // hedgehog-like structures of zinc oxide nanowires

Sustainability aspects increase in the efficiency factor of solar panels and PV elements

PRODUCTS

Structured solar glass

There are now providers of **solar glass** who use a technique whereby the **molten glass** is not dragged through a **liquid tin bath,** but rather through a set of rollers. This results in a pyramid-like structure in the surface, which improves the module's performance by up to 5%.

Sol-gel antireflective coatings

As early as 2005, Merck had succeeded in creating a glass coating based on a **sol-gel technology,** which prevents the reflection of sunlight across a large part of the spectrum and increases the energy output of solar systems by more than 6%. The "sol" consists of tiny silicon dioxide balls measuring 20 to 50 nanometers in diameter, while the thickness of the antireflective coating amounts to around 150 nanometers.

Hedgehog-like structures of zinc oxide nanowires (Source: Swiss Federal Institute of Technology, Zurich)

Hedgehog-like structures of zinc oxide nanowires

At the Swiss Federal Institute of Technology in Zurich, scientists are currently developing a structuring technology to augment photovoltaic systems' efficiency. Through an electrochemical process, tiny hollow polystyrene balls are used to create a hedgehog-like arrangement of zinc oxide nanowires packed tightly together. Scientists assume that this simple production principle will massively increase the surface area of solar cells and thus their effectiveness. In addition, further applications for regular surface structures are expected in the areas of electronics and optoelectronics, for example for **short-wave lasers,** light diodes, and **field emission displays.**

Properties integration of energy-efficient LEDs // semitransparent media facades // three-dimensional light structures with light-conducting fibers

Sustainability aspects low energy consumption // long life

LED MEDIA MATERIALS

LEDs are part of the future of energy-efficient lighting technology. They are extremely long-lasting, need little energy, and can be easily controlled. They are therefore of particular interest for use in media facades and spark the creativity of architects and communication designers in equal measure. One example is the LED installation of the facade of the Esprit showroom in Hong Kong, which was designed and realized by Berlin-based communications agency ART+COM. Here, the designers developed a computer-managed decorative system for the facade sides, which records the goings-on inside the building and replays them on the outside.

Since LEDs can be easily combined with other materials, some manufacturers have integrated traditional materials as a component in their offerings over the last few years. This provides the opportunity for three-dimensional design of light with interaction between material and light source.

PRODUCTS

GKD Mediamesh®

One prominent example is the architectural fabric manufacturer GKD from Düren. There, they have incorporated bright LEDs into **stainless-steel weave** so that it can be used as a surface for information, publicity, and communication in both external and internal spaces. The choice of different types, breadths, and lengths of fabric as well as the **pixel grid** affect the impact the surface has.

IMAGIC WEAVE®

Under the brand name "Imagic Weave," architectural textile manufacturer Haver & Boecker has launched a product that integrates LED technology for **semitransparent media facades.** The textile was used in 2012 for the construction of the Grand Stade Lille Métropole, a football stadium in Villeneuve-d'Ascq, which was created by architects Valode & Pistre in collaboration with Pierre Ferret.

LED media facade at the Esprit showroom in Hong Kong (Source: ART+COM)

IMAGIC WEAVE® in the facade of the Grand Stade Lille Métropole (Source: Haver & Boecker)

ETFE textile architecture with integrated LEDs (Source: Vector Foiltec)

TEXLON® flexipix

Another example is LED-studded **membrane films** made of the light-resistant fluorine-based plastic ethylene tetrafluorethylene (ETFE) for use in textile architecture. The manufacturer has developed a multilayered film cushion with an aluminum frame especially for flexipix. The factory can adapt the LEDs in a straightforward process to incorporate natural light perception.

ETTLIN lux®

Special textile manufacturer ETTLIN has developed a fabric for the implementation of three-dimensional **light structures.** This conducts the light in the **direction of the fibers** to LEDs lying behind it and creates arc-shaped **lines of light.** The optical effect of the light structure changes depending on the angle at which it is viewed. The flexibility of the fabric enhances the design possibilities within the space.

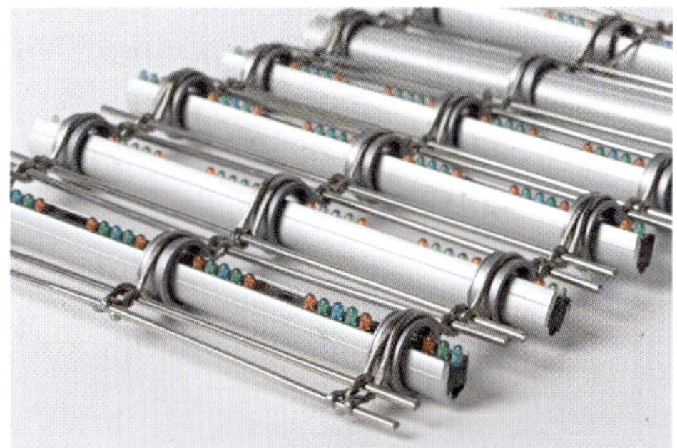

Metal mesh with integrated LEDs
(Source: GKD, photo: Diana Drewes)

Light structure made of ETTLIN lux®
(Source: ETTLIN)

Properties double-sided light emission possible with EL foils // short-wave light // life cycle of 10,000–15,000 hours // arched luminaires possible

Sustainability aspects minimum power consumption // economical use of materials

ELECTROLUMINESCENT MATERIALS

Electroluminescence (EL) describes the light-creation effect of various materials or material compounds once an electric current is passed through them. The effect forms the basis of the way LEDs and OLEDs function. The development of electroluminescent panels, films, and fibers as well as wires has meant the effect is increasingly being used in developments by designers and architects.

MATERIAL CONCEPT AND PROPERTIES

The characteristics of **EL materials** include homogeneous, very short-wave light, minimum power consumption, and low installation depth. Since electroluminescent materials react sensitively to **UV light,** their life cycle is limited by the impact of ultraviolet radiation. The strength of the **electric field** also has an effect on their longevity. Normally a life cycle of far more than 10,000 hours is achieved. Working with shaped EL films makes it possible to create three-dimensional light effects. Double-sided light-emitting EL films can also be used.

EL wires consist of a copper core coated with phosphor and PVC. To ensure light is emitted in all directions, a thin copper wire is wrapped between the phosphor and PVC layers. EL wires are extremely energy efficient.

Structure of EL wires

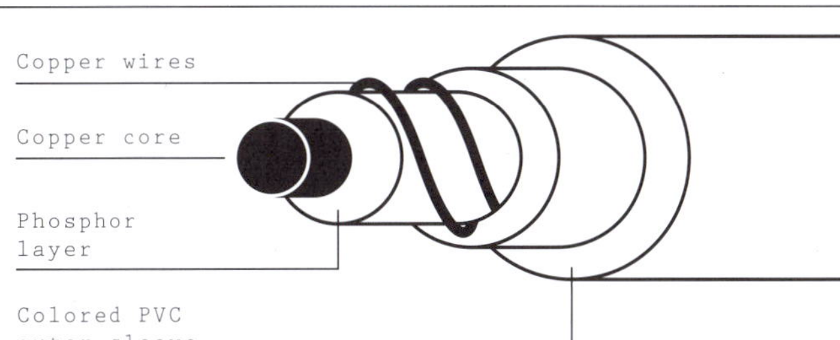

APPLICATION

EL films were first used in vehicle construction in the mid-1960s. Since then, their areas of application have increased considerably. They are in demand wherever low-strength light can offer an advantage. This applies in safety-related contexts as well as for light applications in fashion, design, shop-fitting, and game development.

Spatial installation with EL films (Source: CAAD, Manuel Kretzer)

Structure and functioning of an EL film (Source: modeled after Innoprints)

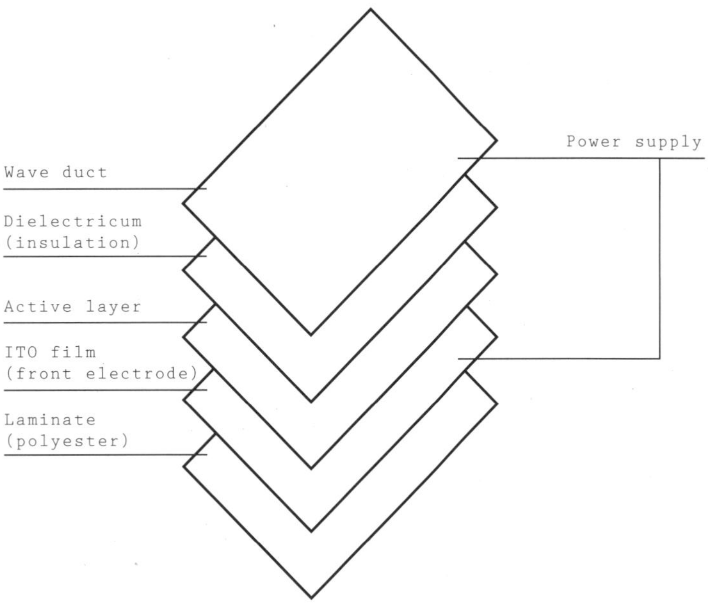

The Perelin luminaires consume very little energy and are extremely resilient due to the special EL film (Design: Markus Becker, photo: Marcus Höhn)

INTERACTIVE LIGHT

Properties light control through human movement // OLED with touch control // touch-free eye commands

Sustainability aspects low energy consumption // control without costly electronics

In interior design, there is currently a strong interest in light installations in which the light responds to the number and movements of people in the space. It is now also possible to obtain OLEDs that can be controlled by touch or by eye movements.

PRODUCTS

Touch-sensitive concrete

One clear example comes from **BlingCrete™** developers Thorsten Klooster and Heike Klussmann. At the University of Kassel, researchers developed **a touch-sensitive concrete surface** that permits the integration of electric switches and thus the formation of entire wall surfaces according to the **touchscreen** principle. Touch-sensitive concrete elements can then turn wall surfaces into control panels for light, ventilation, or heating. Along with the control of light, of technical equipment in the building or of household appliances, elements can also feature a **monitoring function** to measure strain on construction elements or to monitor the **room climate**.

Material animation

As part of the project by the "materiability" research network set up by architect Manuel Kretzer at ETH Zurich, the Swiss Federal Institute of Technology, a moving light installation was created using laser-cut EL films. Due to a variety of wirelessly networked components, this reacts to the movements of people in the space and encourages interaction between them and technology.

Manta Rhei OLED

Due to the possibility of computer-based control of **light surfaces,** OLEDs are considered the next step in the development of the lighting industry. In cooperation with the interactive agency ART + COM, lighting manufacturer Selux has developed a luminaire under the name of "Manta Rhei" with a number of small OLED light surfaces, which floats above the space and continuously and silently changes its appearance in the room. The 1.20 × 2.40 m luminaire is made of 140 wafer-thin **OLED modules** and is suitable for the representation of biological motion sequences in the space. The use of **intelligent controls** ensures that the exposure values of the OLEDs remain stable. Various choreographies can also be presented.

Touch OLED

As early as 2008 Fraunhofer IPMS demonstrated that it was possible to build OLEDs that can be touch-controlled, and now a number of products offering this feature are available on the market. In mid-2012, the Japanese specialists in mobile electronics, NTT DOCOMO presented a transparent OLED solution that could be controlled from both sides through **touch impulses.** The organic display has a diagonal length of 60 mm and a resolution of 320 × 240 pixels. Since both sides function in the form of a **touchscreen,** it is possible to scroll through Internet sites, for example, without the content being hidden.

OLED data glasses with touch-free eye control

At CeBIT 2013, Fraunhofer COMEDD was presented with an innovation award for its **interactive data glasses** with bidirectional **OLED microdisplays.** The user can command the OLEDs touch-free using his or her eyes. So along with a view of the real world, he or she gets additional information according to the principles of **augmented reality,** or AR.

Touch-sensitive concrete for light control (Source: Pat Taylor, Thorsten Klooster, Heike Klussmann)

Manta Rhei (Source: Selux, design: ART+COM)

OLED data glasses (Source: Fraunhofer COMEDD)

Moving light installation using laser-cut EL films and based on interaction between humans and technology (Source: CAAD, Manuel Kretzer)

Properties light production with active organic polymer layer // green-yellow light // low-voltage current sufficient // brightness of 150 candela per m² // relatively long reaction time

Sustainability aspects energy-efficient light production // low energy consumption // straightforward production

Scientists are currently working on a trailblazing development in the area of light technology. With the use of so-called LECs, light surfaces should be economical to create in future and the vision of the illuminating carpet can become reality.

LIGHT-EMITTING ELECTROCHEMICAL CELLS (LEC)

MATERIAL CONCEPT AND PROPERTIES

LECs have a three-ply structure. First, the **light-active organic polymer based on carbon** is printed on a flexible, conductive synthetic film; it is then given a semitransparent synthetic layer made of conductive **PEDOT.** By applying a current of 10 volts, the polymer ions start to migrate between the **plastic electrodes,** which illuminate the films in a green-yellow color. The brightness achieved thus far measures around 150 candela per square meter and can be compared to the brightness of a slightly dimmed notebook display. The low electricity consumption and proven life cycle of thousands of hours are advantages likely to lead to the success of the development of LECs in the market. However, oxygen and humidity do have a negative effect on the lighting power of the LECs.

APPLICATION

The cost-effective manufacture of the **light film** in a rolling pressure process makes mass production possible. This makes applications in interior and fashion design just as realistic as use in packaging. Based on the long reaction time, use in displays is highly unlikely.

Light film with light-active organic polymer based on carbon (Source: LunaLEC)

LEC production in a rolling pressure process (Source: LunaLEC)

Properties biological light // photoproteins for biolasers // low-voltage current sufficient // conditional brightness

Sustainability aspects high energy efficiency // no environmentally harmful emissions

Green-lit beetles, blue iridescent jellyfish, or deep-sea fish with illuminated eyes: light is not only important within our four walls, but also has useful functions in nature in a range of animal species. The glowworm, for example, tries to attract a partner with its glowing green color, and the phenomenon of bioluminescence is also widespread among insects and fungi. Only recently have researchers identified a series of different fungi with self-illuminating capabilities. Over the last few years, bioluminescent light sources have been a key subject of specific research.

BIOLOGICAL LIGHT

BIOLUMINESCENCE PHENOMENA

Dinoflagellate algae

The best-known form of biological light is **marine luminescence.** It is triggered by **microorganisms** such as the one-cell **dinoflagellate algae.** Provoked by an external stimulus, for example a wave movement, the organisms send out light signals. The water is illuminated in a green-blue, with **energy efficiency** at over 90%. In nature, the effect can be experienced, for example, in Phosphorescent Bay near to La Parguera in the southwest of Puerto Rico. With the use of **luminol**, a similar color-light effect can be created by **chemical luminescence** with **hemoglobin.**

Luminescent bacteria

Other natural light effects can be created, among other things, by **luminescent** or **photo bacteria** (for example, **vibrio fischeri**). These live freely in sea water or can be found, for example, on deep-sea anglerfish. The light effects found on foodstuffs (such as **salted herring**) are due to these bacteria. The luminescence of photo bacteria is down to oxidation reactions. Here, the strength of the emitted light depends to a considerable extent on the quality of the organism's habitat. Impurities have a negative effect on the light intensity.

Photoproteins

Another luminescence effect can be found, for example, in the **Aequorea victoria** jellyfish. Here the **photoprotein aequorin** forms the basis for light emission as a consequence of a shift in various energy states. The light emission is due to a return to the initial state through the emission of **photons.** Since the protein is not consumed, the reaction can be reproduced at will. The photoprotein aequorin produces a blue light. The green glimmer is down to the green fluorescent **protein (GFP).**

APPLICATION

The use of biological light for product, fashion, and games design has been the subject of much discussion in recent times. One of the first designers to make use of the bioluminescence of dinoflagellate algae is Nicola Burggraf from Stuttgart. She developed an installation that reacts to the movement of approaching visitors by sending out flashes of light. Like a movement sensor, the installation gives visitors direct feedback on their behavior. Here the light is produced entirely without electricity. The organisms get their energy throughout the day from photosynthesis processes and release it again in the evening, creating light.

The suitability of green iridescent **photoproteins** as a light source for a **biolaser** was proven by scientists in Boston in 2011. To do this, they changed kidney cells through genetic modification, so that they emitted green light when stimulated with pulsating blue light. As part of their research project, Malte Gather and Seok Hyun Yun identified the concentration of the so-called **GFP protein** needed for the creation of **laser light.**

Iridescent bow wave at sea (Source: Wikipedia)

Light installation "Bioluminescent Field" with dinoflagellate algae (Design: Nicola Burggraf)

Bio-Light with luminescent bacteria that have been fed on methane and organic compost (Source: Philips Design)

Green-blue iridescent jellyfish Aequorea victoria (Source: Wikipedia)

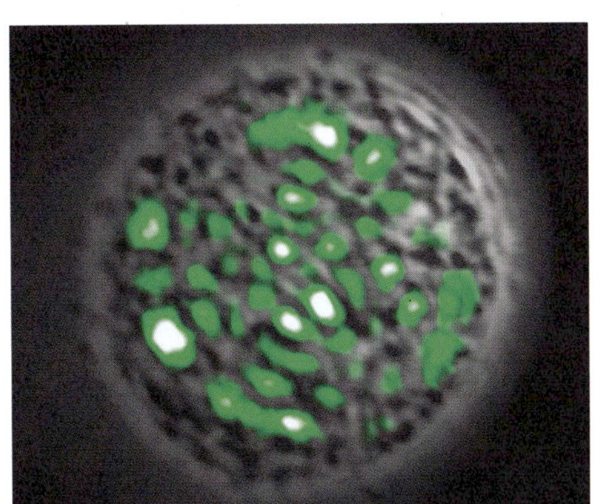

GFP as a light source for a biolaser (Source: MIT)

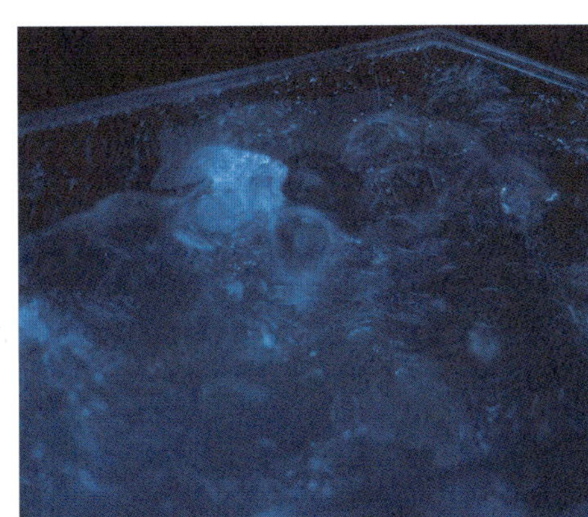

Creation of bioluminescence phenomena through stimulation of dinoflagellate algae with a loudspeaker (Source: Nicola Burggraf)

ENERGY-GENERATING MATERIALS AND INNOVATIVE INSULANTS

07

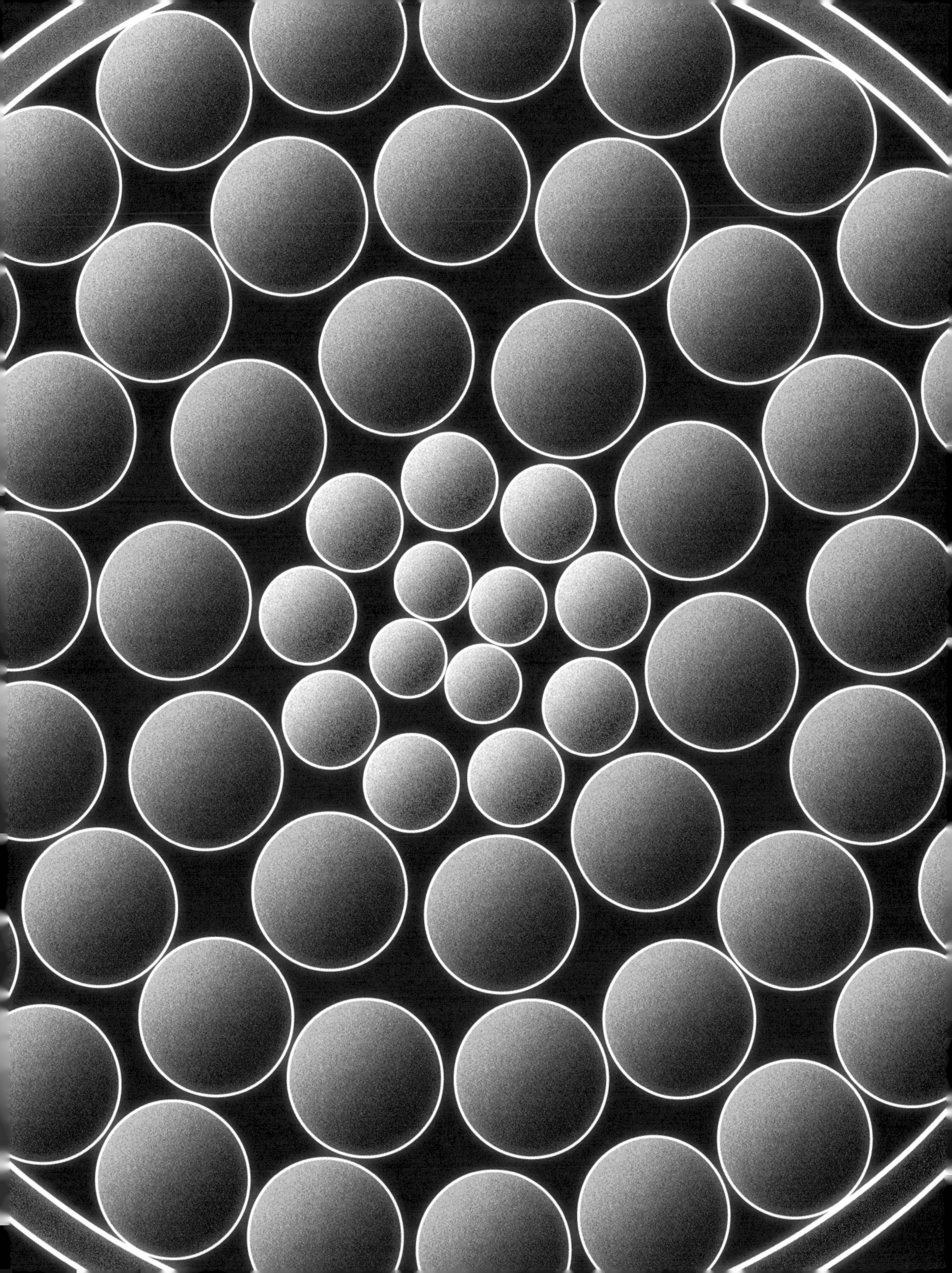

Whether it's facades with organic dye-sensitized solar cells, piezomaterials for the recovery of mechanical energy, or bioadaptive algae facades for energy production in modern architecture: the German federal government's decision for a wide-scale energy reform has triggered a wave of innovation that will render energy production more small-scale and less centralized. The latest price increases among energy providers and the realization that many resources are no longer available in sufficient quantities have made the energy factor a crucial argument when it comes to selling. The energy technology market therefore incorporates not only the systems and components for decentralized energy provision, but also technologies to reduce the demand for energy and measures for increasing energy and material efficiency.

In connection with this reorientation in the energy market, the idea of energy-based materials and technologies is attracting increasing interest from designers and architects. They are the ones who can influence the use of resource and energy-efficient materials and stimulate the development of regenerative energy systems for private use with their future scenarios. This applies particularly to new mobility concepts, such as the use of electric drive and hydrogen fuel cell technology, as well as energy-efficient lights and energy-self-sufficient products based on printed organic electronics and the use of small-scale energy systems. Innovative insulation materials also provide architects with new options for the reduction of heat loss. Likewise, the use of natural insulation solutions with good heat storage capacity offers protection from the heat in summer and thus a reduction in the expense of air conditioning.

171

ENERGY-GENERATING MATERIALS AND INNOVATIVE INSULANTS

Properties printing of active surfaces and electronic switches // light and energy production possible // integration of displays, RFID tags, sensors, or data storage devices // generation of optical structures

Sustainability aspects efficient use of materials // straightforward production // possibility of energy production // energy-efficient light surfaces

PRINTED ELECTRONICS

Given the advanced development of functional liquids and corresponding print technologies, it is now possible to print electronic switches on paper or film, creating energy-efficient light surfaces and displays, electronic components like sensors and data storage devices, and switchable mirrors or solar panels. The research area of polytronics thus has the potential to radically change the use of products and architectural structures and even to make energy-self-sufficient solutions a real possibility. A number of promising offerings have appeared on the market for the implementation of printed electronics:

PRODUCTS

livilux®

Under the brand name livilux®, chemicals corporation Merck offers a wide range of dyes for the realization of organic light-emitting diodes (**OLEDs**). These permit energy-efficient displays to be created with a picture sharpness unaffected by the angle. The livilux® range includes small molecules for **vacuum processes** as well as soluble material systems for print technologies.

PolyTC®

PolyTC® foils consist of thin metal layers on plastic films (usually polyester – PET), which are structured on a micrometer scale and manufactured in **roll-to-roll** processes. They are ideally suited to areas of application where a high level of transparency and high electric surface conductivity is required. In many applications they are a replacement for the widespread **indium tin oxide (ITO) films.** PolyTC® films boast huge potential for use in touch sensors, touch elements, as base electrodes for OLEDs, **organic photovoltaics** (OPV), transparent electrodes, ultrathin **heating elements,** or as flexible circuit boards in specialized areas.

PowerCoat™

This cellulose substrate was developed for the coating of paper surfaces for varied use in printed electronics. PowerCoating produces extremely smooth yet flexible surfaces with improved bonding for colors combined with a reduced tendency for adsorption. The layer system is based entirely on **cellulose** and is therefore biodegradable and suitable for recycling. The increased thermal stability makes economical processing in a **roll-to-roll** procedure possible. PowerCoat™ also permits the incorporation of intelligent functionalities of **RFID tags** in disposable packaging, such as electric switches for illumination purposes and displays, resistors, capacitors, sensors, or **batteries.**

Printed electronics with roll-to-roll process (Source: PolyIC)

ThinFilm™

The Oslo-based developers of ThinFilm™ brought to the market a flexible sticker that can be printed with a **ferroelectric polymer** using conventional print technology. This way, films can be created with integrated displays, sensors, or **data storage devices.** The **memory labels** are particularly significant for **intelligent packaging,** medical devices, ID cards, or interactive games. They can be overwritten, which makes them suitable, most significantly, for the **"Internet of Things,"** a pioneering logistics system from Fraunhofer IML. The company already has storage devices in 4×4 (16 Bit), 5×5 (25 Bit), and 6×6 (36 Bit) formats in its range.

Printed switches for radio-frequency identification (RFID) tags (Source: PolyIC)

The basic materials for printed electronics are functional, electrically conductive, or semiconductive polymers (Source: PolyIC)

LUXeXcel™

The process permits the three-dimensional printing of optically smooth or textured structures, without the need for subsequent processing. Using an inkjet printer, the relevant geometry is created on the basis of CAD data, layer by layer, with a **photopolymer** that sets under UV light. No costly molding tools are required for this, consequently users of optical products can produce **printed optics** individually in batches of one.

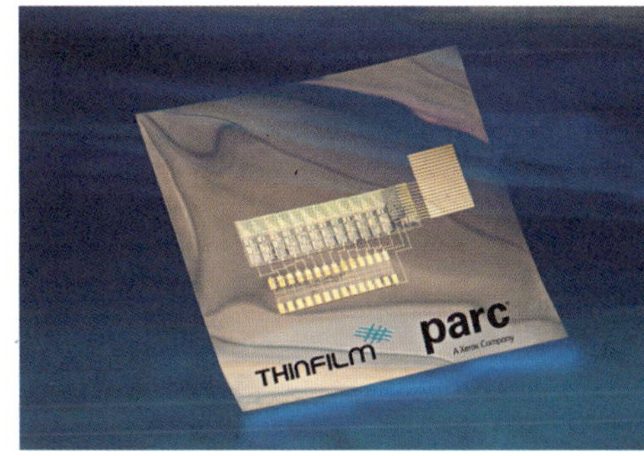

Flexible sticker as data storage medium (Source: ThinFilm™)

Comparison of substrates characterized by irregularities (Source: PowerCoat™)

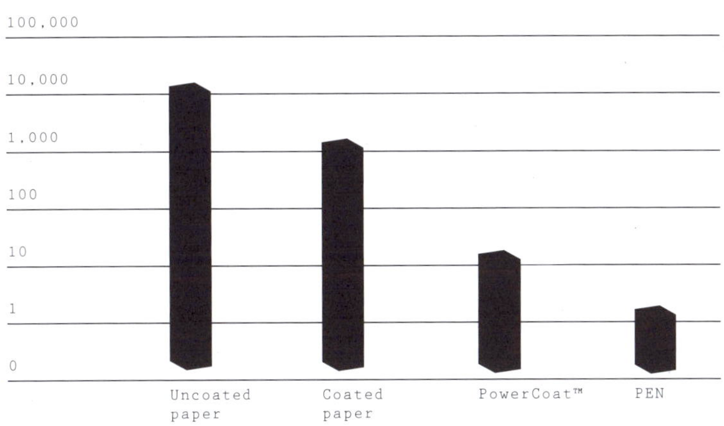

Printed optic lenses (Source: LUXeXcel™)

Properties displays with high energy efficiency // microcapsules with white and black particles with differing charges // applying an electric current leads to a shift in charge // suitable for mobile phones, credit cards, film displays

Sustainability aspects high energy efficiency // simple structure

ELECTROPHORETIC INK (E INK)

Over the last few years, so-called E Ink has become known for its high level of energy efficiency in particular. With the use of E Ink, it is possible to create displays that require 97% less energy than conventional LCD displays.

MATERIAL CONCEPT AND PROPERTIES

E Ink consists of millions of tiny microcapsules with a diameter similar to that of a human hair. Each **microcapsule** contains positively charged white particles and negatively charged black particles, which are suspended in a clear fluid. When an electric current is applied, the particles move to the top of the microcapsule, where they become visible to the user. During the production process, the microcapsules are applied to a synthetic film, which can be cut in various formats and forms. The individual formats make up the segments of a display, which can be commanded in various ways and thus appear either black or white.

APPLICATION

E Ink permits the creation of displays and advertisements with high-resolution images and texts. The latter are achieved due to a **thin film transistor** (TFT) array made up of points or pixels. In product development, designers can work with either the display film or with finished modules. E Ink has already been used successfully in mobile phones, **film displays,** and credit cards.

Functioning of E Ink (Source: E-Ink)

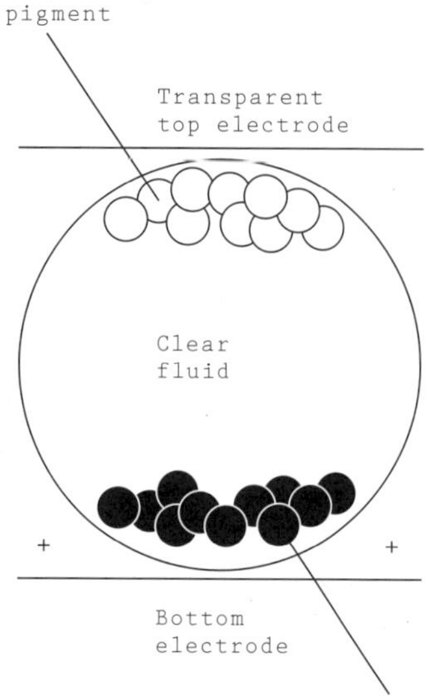

ORGANIC PHOTOVOLTAICS (OPV)

Properties energy conversion with polymer semiconductors // OPV cells achieve efficiency of 12% // angle-independent light absorption // functional even in internal spaces

Sustainability aspects no hazardous materials required // efficient use of materials // straightforward production

The potential of OPV has long been of interest, but until now the offering has lacked stability and the energy conversion has not been efficient enough. This is all set to change. Some manufacturers have set themselves the objective of bringing to the market systems that convert the light entering them to electrical energy and which have a life cycle of significantly more than 10 years. In January 2013, Heliatek® in Dresden announced the development of an OPV cell with efficiency of 12%, which represented the best level so far. Manufacturers predict that in the long term, OPV will get more efficient and cheaper than its counterpart based on silicon.

MATERIAL CONCEPT AND PROPERTIES

OPVs are based on **organic semiconductor materials,** i.e., **hydrocarbon compounds** (so-called conjugated polymers) as the primary **absorbing material.** The application of the individual layers of the sandwich structure can be carried out through **silkscreen, roll-to-roll,** or **spinning processes.** Work with cost-intensive **evaporation** or **vacuum techniques** is not necessary. In addition, absorber layer thicknesses of just 100 nanometers mean less material is needed. In Heliatek's product, just one gram of material is needed for the production of a cell surface of one square meter. When working with film-like support materials, the essential advantages of OPV are the flexibility of the cells and their light-permeable transparency. OPVs absorb the light regardless of the angle and convert it into electricity even on cloudy days. They also have potential for **energy conversion** in internal spaces.

APPLICATION AND PROCESSING

In the long term, the main application will be in the construction industry, because their transparency means that OPVs can be integrated as thin synthetic layers in roofs, windows, or facades. In addition, they will ensure mobile use of electrical devices. As their efficiency increases, OPVs should be suitable for charging notebooks. For outdoor applications, the possibility of manufacturing folding and flexible OPV films is particularly relevant. It is also conceivable that transparent film could be used on car roofs to produce the required electricity for air-conditioning systems. Given their strong light absorption, they can be vapor-deposited in thin layers and printed with conventional techniques.

Smart ForVision electric vehicle with roof made using OPVs (Source: Daimler, BASF)

OPV use in architecture (Source: BELECTRIC®)

PRODUCTS

lisicon®

Under the name lisicon, a series of **printable polymers** for the cost-efficient manufacturing of OPVs has been launched in ready-for-use formulations, which can be applied using conventional printing techniques. These include the typical inkjet roll process as well as **gravure** or **flexo printing.** Developers also think that coating using **spin printing techniques** is a possibility. These production options mean that **organic electronics** are growing in significance.

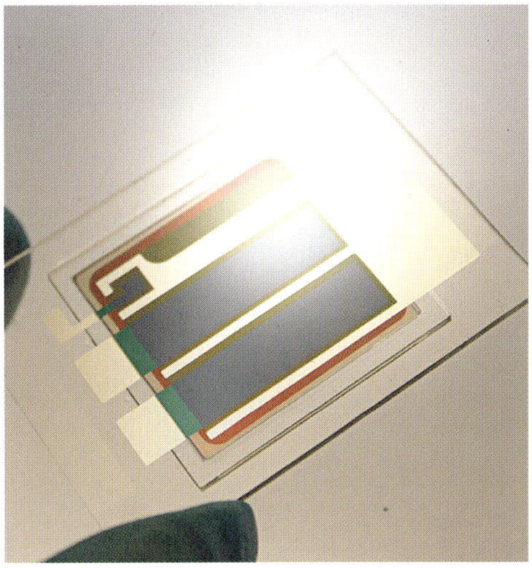

Two Heliatek world record cells with 12% efficiency on an active surface of 1.1 cm² for each (Source: Heliatek)

Structure of an OPV (Source: University of Oldenburg)

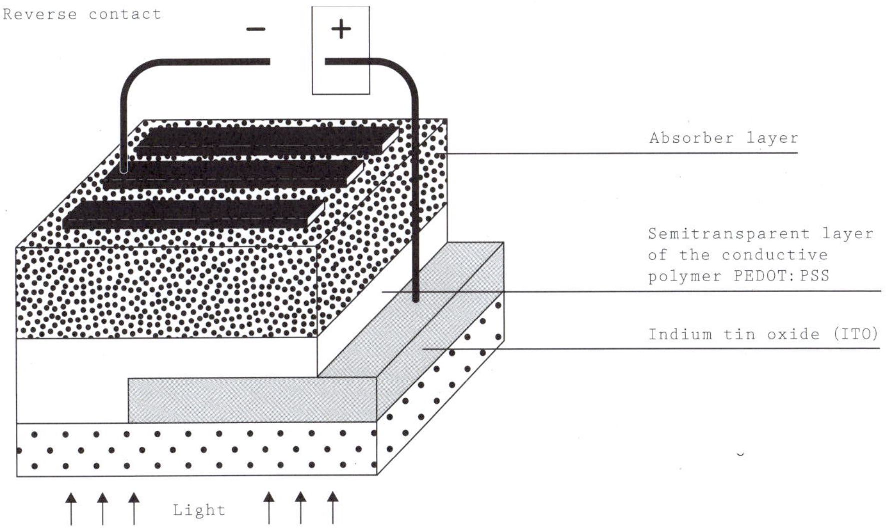

BELECTRIC®

Over the last few years, certain essential improvements have been brought about in OPV which have made transparent and color-formatted systems of interest for use in flexible architecture or in sport and outdoor systems. BELECTRIC® aims to establish itself on the market with new offerings for the printing of OPVs.

Heliatek®

The development of Heliatek® cells with their current highest possible efficiency combines two patented **absorber materials** and particularly efficient energy utilization over a higher **photoelectric voltage.** A series of measurements taken by an accredited testing institute confirmed that, due to the above-average efficiency even with a low level of light and high temperatures, the **efficiency factor** of 12% and **cell efficiency** of 14 – 15% are comparable to conventional crystal silicon modules and **thin-film photovoltaics.**

Properties solar cells based on an electrochemical reaction // functions with organic dyestuffs such as chlorophyll or grape juice // longevity and efficiency factor still low

Sustainability aspects energy production without emissions // dyestuffs based on renewable raw materials possible // recyclable

DYE-SENSITIZED SOLAR CELLS

Given their straightforward construction and the low costs of the necessary components, dye-sensitized solar cells (DSSC or DYSC) are considered a low-cost solution for the future with a broad range of applications. The self-build format is enjoying increasing popularity among designers and architects. As early as 1991, chemist Michael Grätzel developed the principle, based on photosynthesis, which is why the construction became known as the "Grätzel cell."

MATERIAL CONCEPT AND PROPERTIES

The conversion of solar into electrical energy takes place on the basis of an electrochemical reaction with organic dyestuffs such as **chlorophyll** or **grape juice.** The construction requires two glass plates in addition to the dyestuff. One is coated with **porous titanium dioxide** and covered with the dyestuff. The other glass plate is given a graphite coating and functions as a negative electrode. An electrolyte (such as **common salt**) is applied between the two. When sunlight enters, the dyestuff is stimulated and releases electrons to the titanium dioxide. An electric potential gradient is formed.

DSSCs can be produced with no toxic emissions as freely obtainable components. They are largely recyclable and environmentally friendly to run. Dye-sensitized semiconductor materials make transparent and flexible support systems possible. As a result they can be used in vehicles, in furniture construction, for clothing, or for building facades. Numerous designers and architects have already experimented with them, and scientists are working intensively on improving efficiency and longevity above all.

PRODUCTS

DysCrete

In the department headed by artist Heike Klussman at the University of Kassel, research is currently being carried out into straightforward processes for the manufacture of DSSCs. The main objectives are the experimental development of new approaches for solar cells integrated into buildings as well as new solutions for mobile systems. The researchers are aiming predominantly for the combination of DSSC and concrete. Through the targeted recycling of sun protection glass from demolition and glass breakage, the researchers have already been able to produce prototypes for less than €5 per square meter.

Structure of the DSSC (Source: Thorsten Klooster)

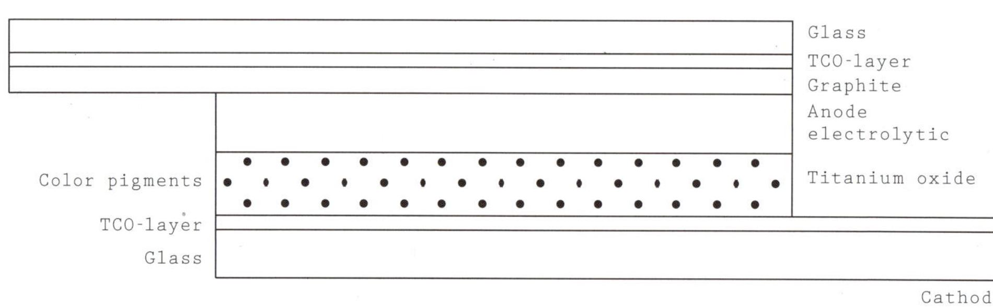

Active dyestuff (Source: Heike Klussmann, Thorsten Klooster & Negar Jahadi Rafigh, photo: Pat Taylor)

Dye-sensitized, energy-producing concrete (Source: Heike Klussmann, Thorsten Klooster & Negar Jahadi Rafigh, photo: Pat Taylor)

ENERGETIC TEXTILES

Properties conversion of body movement into energy // integration of flexible DSSCs in textiles // operation of sensors for body monitoring

Sustainability aspects operation of energy-self-sufficient systems in textile applications

In future there will be an increasing focus on energy-self-sufficient products capable of deriving energy from environmental influences. For energy-harvesting in the textiles sphere, along with dye-sensitized and thin-film solar cells, materials with piezoelectric properties are particularly suitable.

PRODUCTS

Piezoelectric textiles

Piezoelectric materials can convert pressure and gentle expansions into electrical impulses. In this context, **piezo crystals** usually come to mind, which have long been used to take on a variety of functions in regulatory and control technology. However, there are also synthetics with piezoelectric properties that can be integrated into textiles for the purpose of energy generation. These are currently being investigated in the European research project PIEZOTEX at the Fraunhofer IIS. One of the subjects of this research is the development of a **PVDF yarn** to incorporate electric components into textiles. Applications could include sports clothing, for example, with integrated sensors to monitor body functions, or textiles that can support work processes.

Fiber-based DSSCs

In the TEXSOLAR project at TITV Greiz, scientists have succeeded for the first time in achieving energy conversion directly on **dye-modified textiles.** If this could be achieved under production conditions, it would be possible to manufacture cost-effective, highly stable and self-sufficient textile sensors and highly flexible electronic components for use in medicine and technology. The textile DSSCs are suitable for microsystems with small solar surfaces for supplying energy to interactive textile components with energy consumption of up to 100 microwatts.

APPLICATION

In the near future, energy-generating textiles will be used, for example, for sports clothing, which can monitor body functions by means of integrated sensors. Furthermore, there is talk of textiles for work clothing, which will support work processes through to the integration of minor support systems.

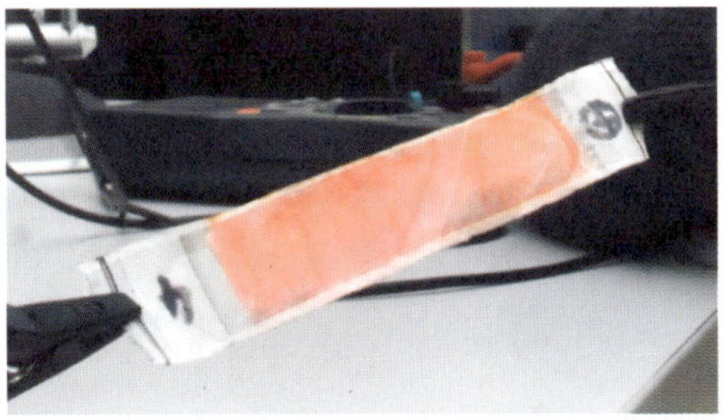

Textile DSSC
(Source: TITV Greiz)

Miniature voltage converter for piezoelectric textiles
(Source: Fraunhofer IIS)

Proteins, viruses, and bacteria are all terms attributed to chemistry, biology, and medicine. The fact that they can also play a part in an energy context is astonishing and shows the extent to which scientists are making use of natural resources.

Properties energy generation using viruses, bacteria, and protein // functioning without rare metals // potential also for sensors, computer chips, or transistors made of organic materials

Sustainability aspects energy production without emissions // based on renewable raw materials

BIOLOGICAL ENERGY

MATERIAL CONCEPT AND PROPERTIES

Biohybrid cells based on spinach protein

In the USA, research is currently being carried out into **solar biohybrid cells** that use a photosynthetic **protein** made from **spinach plants.** In their research, the scientists have worked entirely without rare earth metals and already achieve 850 microamperes per square centimeter with a current of 0.3 volts. The research team is not only relying on spinach as the source for **solar protein,** however, proteins from kudzu are also being investigated. In both cases a protein solution was applied to a silicon semiconductor wafer and the water was then evaporated in a vacuum. The result is a firm protein layer, the optimum thickness of which is around one micrometer. Scientists believe that biohybrid **solar cells** should be able to compete with conventional PV cells within three years at most.

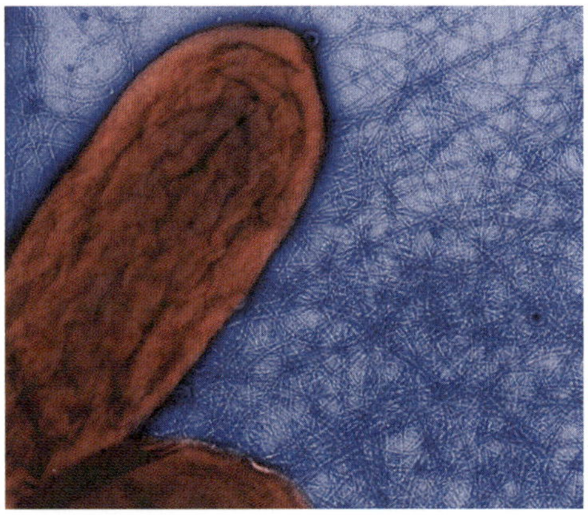

Bacillar microbes under the microscope (Source: Anna Klimes, Ernie Carbone, University of Massachusetts)

Electric generator based on viruses

In 2012, researchers at the University of California in Berkeley managed to produce electric power with a generator that uses **viruses.** The so-called M13 **bacteriophages,** which are bacillar and not harmful to humans, were arranged in 20 positions between two **gold electrodes** in such a way that they change position and electrical charge under pressure and thus produce electricity. The scientists have already been able to produce 6 nanoamperes of power and 400 millivolts in voltage with a **viral generator** of just 1 cm^2 in size. This energy yield would be sufficient to supply a small-format LCD display with electricity. Integrated into the sole of a shoe, the viral generator would produce sufficient energy, similarly to the **piezo effect,** to power a mobile phone, for example.

Biohybrid cells based on spinach protein (Source: Amrutur Anilkumar, Vanderbilt University, Nashville)

Electrically conductive biofilms

Bacillar **bacteria** made of fine **protein threads** may also be important in the future for electronic components, as an international team of scientists from the University of Massachusetts has proven that **biofilms** made of microbial **nanothreads** would conduct electricity in precisely the same way as those made of conventional materials. The researchers were able to influence the conductivity of the **microbes "geobacter sulfurreducens"** significantly, for example through temperature change and the variation of **gene activity.** Sensors, computer chips, or transistors could thus be produced from organic and nontoxic materials in future.

In future, the production of solar cells could become much cheaper, as American and German researchers have succeeded in printing solar cells on simple paper. This way, in the future, it would be possible to create solar cells with a record yield in terms of the ratio of electrical output to mass.

Properties printing solar cells on paper // record yield of electrical output in relation to mass // 250 tiny solar cells on 49 cm² // efficiency factor of 5% targeted in the long term

Sustainability aspects energy production with simple starting materials // recyclable // cost-effective

SOLAR PAPER

MATERIAL CONCEPT AND PROPERTIES

At MIT in Massachusetts scientists are using a **vacuum vapor-deposit process** with the integration of a layer made of light-active thiophene polymer to introduce the five layers needed for the paper. A surface of just 49 cm² accommodates a total of 250 small solar cells, which supply output voltage of between 26 and 50 volts when illuminated with a halogen lamp. The output of the folding solar paper was sufficient to supply a flat laptop monitor with energy. The efficiency factor of the solar paper is currently rated at a good 1%. In the long term, however, values of over 5% should be possible. What is particularly striking is the material's robustness. Despite frequent rolls, folds, and creases, the **solar paper** reliably supplies power for a duration of 6,000 hours. It can be mixed with normal recycled paper and disposed of in the usual fashion.

In another development at the Chemnitz University of Technology, **electrically conductive colors** were processed with conventional print technologies and **paper solar cells** created with an efficiency factor of 1.3%. This is set to be increased to 8% as a laboratory benchmark in the medium term.

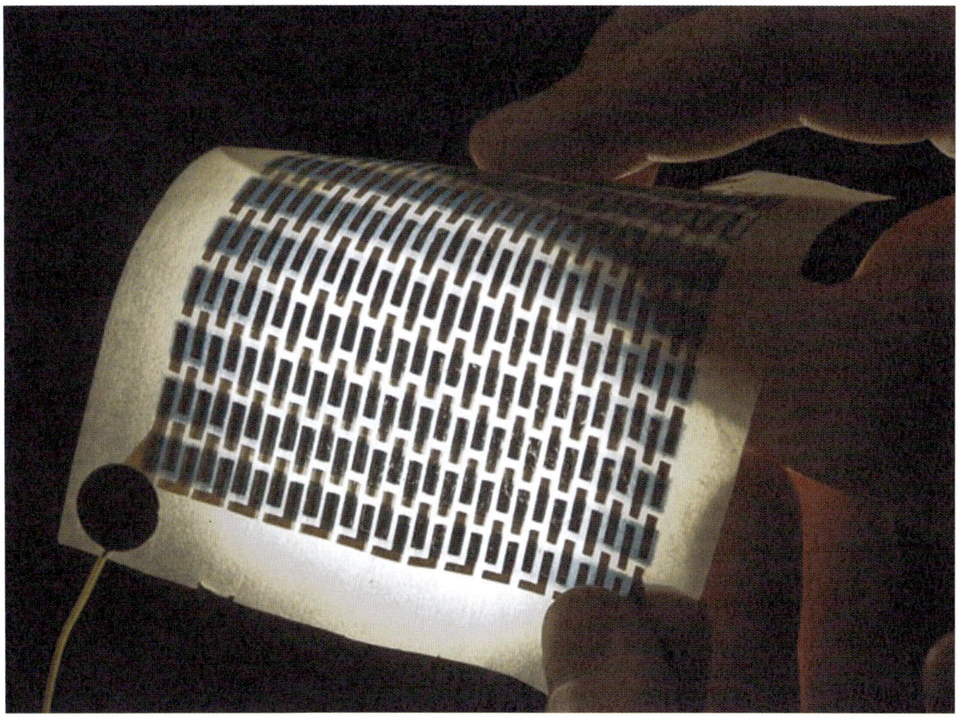

Solar paper
(Photo: Institute for Printing and Media Technology at Chemnitz University of Technology)

Solar paper being tested
(Source: TU Chemnitz)

APPLICATION AND PROCESSING

As well as being a power supply for mobile devices, in future solar paper could be used in **intelligent packaging.** Another possibility might be wallpaper for powering room lighting with light-emitting diodes. The fact that these future materials are simple to use and cost-effective in production makes them suitable for use in developing countries.

Properties energy recovery from warm waste gases with thermoelectric plastics // temperature gradient of 30 °C required on material front and reverse sides // potential for intelligent clothing and monitoring systems

Sustainability aspects energy recovery // battery replacement for mobile applications

THERMOELECTRIC PLASTICS

The increase in energyefficiency is one of the leading topics in the current energy debate. In industrial plants or at the exhausts of vehicles, hot gases are pumped out that generally have no further use, even though the energy they contain could be used. One possibility is working with thermoelectric generators for energy recovery. Solutions based on bismuth telluride alloys are the most widespread. Now scientists are harnessing the potential of thermoelectric plastics.

MATERIAL CONCEPT AND PROPERTIES

Thermoelectric materials are substances that are able to produce electrical power with low temperature fluctuations on the front and reverse of the material. Researchers at the University of Linköping, Sweden, report that they have successfully developed a plastic layer with thermoelectric properties from the polymer **polyethylenedioxythiophene** (**PEDOT**). With a temperature gradient of 30 °C, the coating has already produced enough electrical current to supply small switch circuits with sufficient energy. The heat conductivity of the plastic is reduced. During the production process, the polymer was mixed with a solution that contains iron and the liquid material was applied to a glass plate. The yield still amounts to just a few microwatts, but researchers are currently working on improving the efficiency factor.

APPLICATION

Thermoelectric materials offer the possibility of producing small amounts of energy for mobile applications and could therefore replace **battery systems.** New applications are also possible, for example **intelligent clothing**, **sensor-controlled dosage of active agents** in medical technology, or the monitoring of technical systems on the outer surface of aircraft, on brake discs, or on vehicle axles.

Vehicle exhaust
(Photo: Diana Drewes)

The combination of thermoelectric and metallic materials makes it possible to print structures that can be used as thermoelectric generators (Source: Fraunhofer IFAM)

Properties natural fiber insulation // Neptune grass has highest heat storage capacity of any known insulant // sheep's wool can absorb odors and pollutants

Sustainability aspects good heat insulation of natural fibers with high heat storage capacity // based on renewable raw materials

NATURAL INSULANTS WITH GOOD HEAT STORAGE CAPACITY

Comparison of heat storage capacities of various insulating materials
(Source: NeptuTherm®)

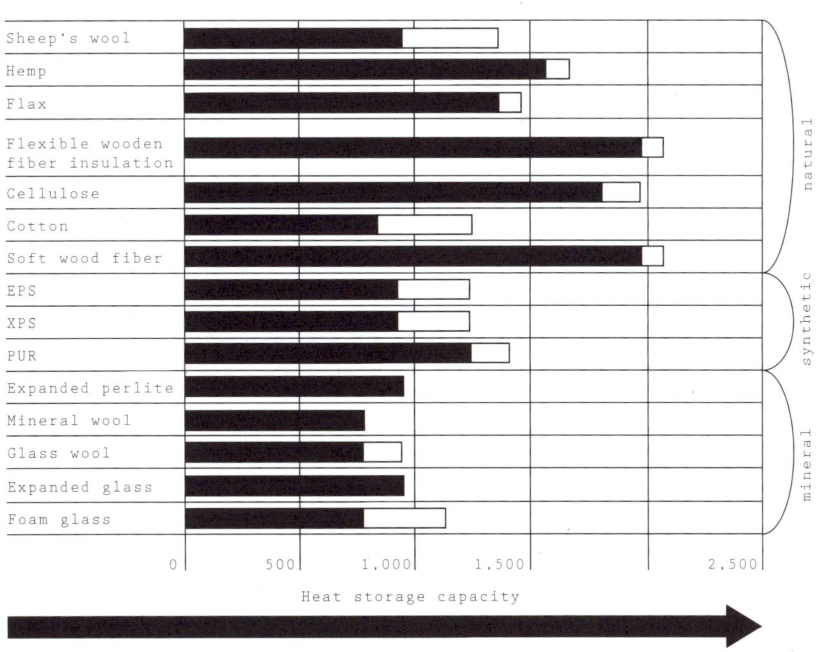

Neptuflex insulating mat
(Source: NeptuTherm®)

Insulating materials made of sheep's wool
(Source: Baur Vliesstoffe GmbH)

Although the properties of heat insulation based on natural insulating products are inferior to those of inorganic materials such as glass or stone wool or based on PUR, their use is favorable from other points of view, such as saving resources and primary energy use. This predominantly concerns insulating products that come from recycled materials or from animal or plant-based substances. In addition, natural insulating materials offer advantages for heat protection in summer due to higher heat storage capacities.

PRODUCTS

Neptune grass

Neptune grass is the term used for the flotsam of **seaweed plants.** These play a significant role in the global ecosystem, since one hectare of the marine plant can convert more carbon dioxide into oxygen than the same area of a tropical rainforest. In the Mediterranean, this fibrous material rolls up into balls. These are collected, unraveled, and marketed as a heat insulation material (conductivity: 0.043 watts per meter per Kelvin) with **natural fire protection** and outstanding heat storage capacity. The manufacturer NeptuTherm® was able to prove this in 2012 with a value of 2.502 joules per kilogram Kelvin. Neptune grass thus surpasses the **heat storage capacity** of all known insulating materials and is almost 20% better than wooden fiber insulation.

Sheep's wool

Animal **hair** has the added property of absorbing moisture and releasing it into the environment where necessary. This means insulating materials made of **sheep's wool** have a positive influence on the **room climate** and help to keep the moisture in the air at a healthy level. With its high **bioreactivity,** wool can also absorb odors and pollutants and thus neutralize them to a certain extent.

AERO-INSULANTS

Properties ceramic sponge structure with air proportion of 95-99% // aerogel integrated construction material with heat conductivity of 0.016-0.019 W/mK // 70% slimmer construction possible // particularly suitable for internal spaces // flexible aerogel including for piping

Sustainability aspects outstanding insulation properties // material-efficient insulation solutions

Their particular properties for heat insulation make aerogels an outstanding example of nanotechnology. The reason for this is a ceramic sponge structure, which contains between 95 and 99% air. Now, more and more companies are incorporating the material into their products and developing solutions for applications in new areas.

PRODUCTS

Sto-Aevero

Through the integration of **aerogels,** the building materials manufacturer Sto has developed an internal insulation solution with **heat conductivity** of just 0.016 W/mK and a thickness of just 10 to 40 mm. Sto-Aevero is suitable not only for **surface insulation,** but can also be used to insulate **window reveals** and radiator recesses.

Aerowolle®

This is the name given to an insulant now available on the market, whose combination of mineral wool and aerogel gives it outstanding **insulation properties.** Its heat conductivity stands at 0.019 watts per meter per Kelvin, which permits constructions 70% slimmer than conventional solutions.

AeroClay®

AeroClay® was discovered at the Case Western Reserve University in Cleveland during an attempt to use clay as a basis for the manufacture of aerogel. After a dispersion of water and **clay** had been freeze-dried, the result was an extremely fragile structure, to which were added **biodegradable polymers** as a binding agent to improve its mechanical properties. AeroClay® can be processed to produce different formats, regains its shape well and is suitable for various applications as an insulant, packaging material, and for the absorption of oil.

Flexible aerogel
(Source: NASA)

Flexible aerogel

This redevelopment by NASA transferred the insulation properties of the fragile aerogel variants into a material with highly flexible properties. In this way, in future it will be possible to use the aerogel properties for high-insulation clothing, fridge walls, and piping.

Enova® aerogel

Under the name of Enova® aerogel, aerogel manufacturer Cabot has developed a new additive for paint which can be used to enhance insulation properties of metals both for heat and cold. This is achieved through a water-based coating. Depending on the formula, a thermal conductivity of 30 – 50 W/mK can be achieved, providing an insulation between seven and ten times better than ordinary paint. Areas of application include cold chambers, metal roofs, and rooms with particular temperature requirements.

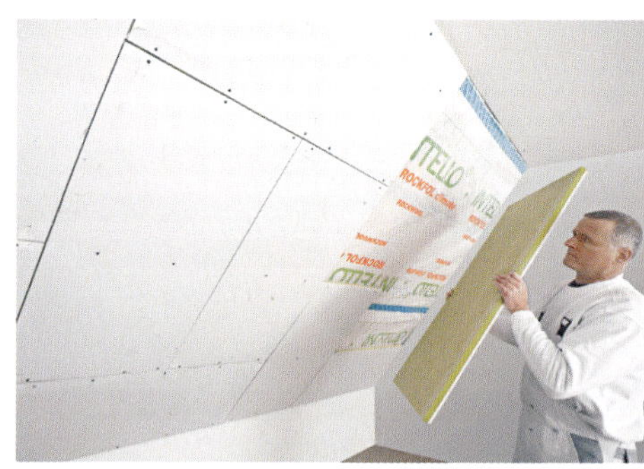

Insulating panel made of Aerowolle®
(Source: Rockwool)

Sto-Aevero
(Source: Sto AG)

Researchers from the Institute of Textile Technology and Process Engineering Denkendorf (iTV) are hoping to increase the insulation properties of construction materials by using the structure and format of polar bear fur.

Properties insulants modeled on the fur of polar bears // hollow fibers direct solar radiation to a black absorption surface // textile-based solar panels for energy-self-sufficient pavilion successfully tested // able to function in summer and winter

Sustainability aspects highly efficient heat insulation // improvement in energy balance // energy-self-sufficient buildings possible

INSULATION SYSTEM MODELED ON THE POLAR BEAR

MATERIAL CONCEPT AND PROPERTIES

Thanks to **transparent hollow fibers,** sunlight is directed onto the animal's black skin, which absorbs heat radiation, and the loss of heat is prevented by the incorporation of minute air spaces. This way, polar bears are able to survive even the very lowest temperatures. This structure inspired scientists at iTV Denkendorf to develop a transparent heat insulation system. This has a block film as a basis, on which a **fibrous fabric** is laid, which is then covered by another, this time transparent, film. Sunlight can penetrate to the black film and heat it up. Heat radiation cannot escape, since it is reflected by a **nanostructured layer** on the upper film and contained in the space in between. Water or air can flow through the fabric and be heated.

APPLICATION

The scientists are hoping that **textile-based solar panels** will have numerous possibilities for application in solar thermal solutions in the construction industry. With the insulation material modeled on the polar bear, it is possible to achieve the standard of a **passive house.** January 2013 saw the opening of an **energy-self-sufficient pavilion** made using **textile membranes** incorporating the **polar bear insulation system.** The use of flexible solar panels ensures a heat supply in summer and winter alike.

Energy concept of a "polar bear building" (Source: iTV Denkendorf)

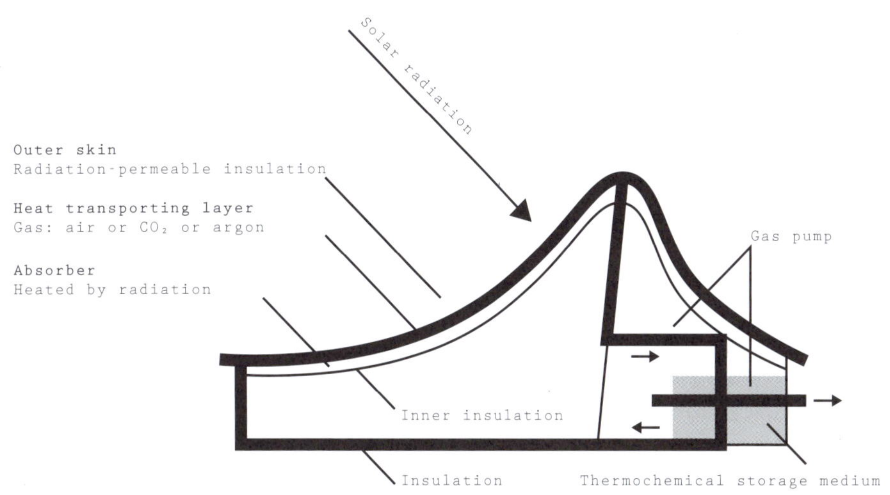

"Textile"-based thermal solar panels modeled on polar bear fur (Source: iTV Denkendorf)

Polar bear pavilion (Source: iTV Denkendorf)

Properties high-performance electricity conductor for efficient energy transport // conductor cables made of temperature-resistant aluminum, aluminum oxide fibres, and molybdenum, tungsten, and carbon composite being researched

Sustainability aspects Material efficiency owing to improved conductivity

HIGH-PERFORMANCE MATERIALS FOR ENERGY CONDUCTORS

Due to the shift in energy policy, new transmission systems are required to transport electrical energy efficiently throughout the country. The German Energy Agency (dena) has calculated that new transmission lines with a total length of 3,600 kilometers will be needed by 2020, the realization of which is questionable due to protests by local inhabitants. An alternative would be the use of highly efficient energy conductors on the old routes.

MATERIAL CONCEPT AND PROPERTIES

High-temperature-resistant aluminum

Under discussion, for example, is the use of conductor cables made of temperature-resistant aluminum. Standards limit the maximum heating by electricity conduction to a maximum temperature of 80 °C. The problem with conventional **aluminum-steel cables** is that they stretch considerably at high temperatures. The cables would sag and therefore require an expensive increase in pylon height. Through the incorporation of **zirconium**, **temperature-resistant aluminum** (**TAL**) can resist temperatures of 150 °C and can transport around 50% more electricity. This would avoid the construction of 1,700 kilometers of lines.

Aluminum oxide fibers with aluminum wire sheathing

Alternatives would be electricity conductors made using a combination of tens of thousands of ultrahigh-strength **aluminum oxide fibers with aluminum wire sheathing** (**ACCR**) or with a **carbon fiber core** (**ACCC**), which is less inclined to stretch when heated. The use of such materials permits an increase in capacity of 100% on current conductor systems.

Superconductors

ThyssenKrupp announced another alternative high-temperature conductor made of a mixture of **molybdenum, tungsten,** and **carbon**. Other so-called superconductors with conductivity up to 10 times that of conventional copper are made from **multifilament ceramics**, which are reinforced with a **silver magnesium alloy.**

Conductor cable 3M™ Aluminum Conductor Composite Reinforced (ACCR)
(Source: 3M)

Properties biomass production with algae facade // solar thermal generation of heat for building services // yield: 15 g dry mass per square meter and day // net energy gain of approx. 4,500 kWh per year // new prospects for light distribution and shading

Sustainability aspects heat and biomass generation on previously unused surfaces // algae produces biomass very quickly // carbon dioxide is absorbed

BUILDING-INTEGRATED PHOTOBIOREACTORS (PBR)

The development of intelligent and functional facade surfaces is one of the current focal points of research in the field of architecture. While in recent years very interesting multimedia facades with particular light effects have been implemented, the notion of the facade as a means of energy generation is gaining in significance. Thus, for the occasion of the 2013 International Building Exhibition in Hamburg's Wilhelmsburg district, bioadaptive facade elements were developed.

FUNCTIONAL PRINCIPLE AND STRUCTURE

Bioreactors are incorporated into the facades in which **microalgae** are cultivated. **Algae** are easy to cultivate and convert sunshine and carbon dioxide into biomass and heat. The vertical **glass louvers** serve as stores for both the resulting biomass and the heat held in the watery liquid. The biomass produced and the heat are circulated and can be extracted when required. While the resulting solar thermal heat can be used for the direct heating of the building, from time to time the **algae biomass** can be removed from the facade and recycled in a biogas plant. From a design perspective, particularly interesting elements are the natural transparency of the algae and their striking green color. This opens up new perspectives in light distribution and shading.

APPLICATION

The first bioadaptive **algae facade** was presented in 2013 – in **Wilhelmsburg** on the **BIQ (Bio-Intelligence Quotient) building** covering a surface of 200 m². The individual **louvers** are 700 mm wide and 2.6 m high, with a filled weight of 200 kilograms. With a yield of 15 grams of dry mass per square meter and day, you can assume a **net energy gain** of around 4,500 kilowatt hours per year with the conversion of biomass into biogas. At the same time, an efficiency factor of up to 80% is achieved.

Facade elements for microalgae cultivation (Source: Arup)

Energy concept of the bioreactor facade (Source: BIQ – The Algae House)

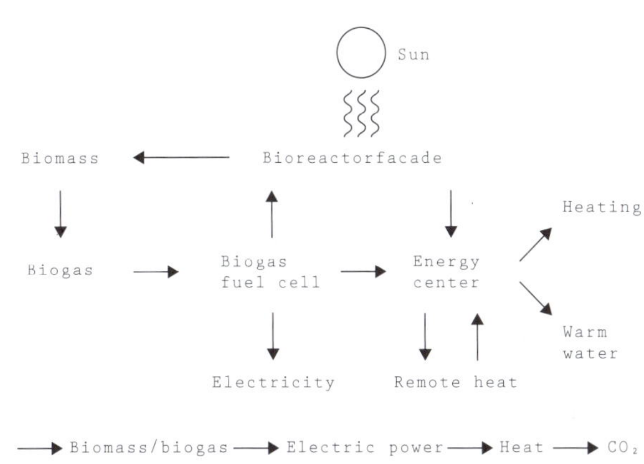

INNOVATIVE AND SUSTAINABLE PRODUCTION PROCESSES

08

Innovations in production technologies aim to lower the use of environmental pollutants, make production processes as efficient as possible, and reduce consumption of resources to a minimum. In particular, generative technologies offer considerable potential here. For in contrast to conventional machining processes such as lathing, drilling, and milling, the additive production process, layer by layer, only requires material where it is needed based on aesthetic criteria and mechanical load. It is precisely in this field that a number of revolutionary techniques have been developed in recent years enabling, for instance, the assembly of components using sand subjected to intense sunlight, or continual generative production on a conveyor belt. With "bioprinting," scientists have even begun working on techniques to print food, and animal or human tissue.

Other developments focus on optimizing materials for considerably longer use, enabling the production of three-dimensional components by means of fibers, or reducing material consumption in joints between plastics and metals. Today it is just as feasible to transform natural cellular and open-pored structures into ceramic materials making efficient use of resources as it is to integrate graphic structures in concrete surfaces. A development by a start-up in Bielefeld shows that different materials in composite structures can be separated by rinsing processes. Using nanoscale surfactants, the company is able to separate the valuable metal layers from the glass surface of a photovoltaic element as well as break up the compounds between metals and plastics in packaging.

INNOVATIVE AND SUSTAINABLE PRODUCTION PROCESSES

Properties fusing of sand particles // focuses the sun's energy // layer-by-layer construction // indentations possible only to a limited degree

Sustainability aspects no additional power sources required // sand available in abundance as the raw material // regional products

SOLAR SINTER RAPID MANUFACTURING THROUGH SUNLIGHT

Laser sintering is one of the most important technologies in the field of generative manufacturing for the production of molded parts. Since the end of the 1990s, it has been used to manufacture high-grade technical prototypes from ceramics, plastics, and metals. Since then, technological developments have made it possible to work with almost any material available in powder form. Above all, this technology has enabled the manufacture of molded metal components for tools (rapid tooling) and even standalone objects (rapid manufacturing). In the Egyptian desert, German designer Markus Kayser demonstrates what the next great step forward in development may look like as he produces glass components using only sand and sunlight.

PROCESS PRINCIPLE

All this is made possible by the Solar Sinter system, which was developed by the designer himself. Sunlight is focused onto a specific point in the **bed of sand** through a lens assembly made up of multiple **Fresnel lenses.** The resultant heat is sufficient to cause the **silicon particles** to bond with one another, turning to glass as they cool. A **solar-powered** 3D printing device, also built by the designer, is tasked with bringing a fresh layer of sand into position as it pushes the previous layer of the object downward. This process is repeated as many times as necessary to produce a finished version of the desired object.

APPLICATION

The **Solar Sinter** enables Markus Kayser to bring high technology to newly industrialized and developing nations using simple and – more importantly – locally available materials, making it possible here too for the benefits of generative manufacturing processes to be exploited and further developed.

Bowls sintered using sunlight
(Source: Markus Kayser)

Solar Sinter in the Egyptian desert
(Source: Markus Kayser)

One approach to encourage a more economic use of materials in the manufacturing world is to incorporate waste materials and recycled scrap into the manufacturing process. Thermoplastic waste materials melt at a low temperature and so are particularly well suited. Dutch designer Dirk Vander Kooij succeeded in using waste materials for generative manufacturing in 2011.

Properties fusing of used plastic materials and wood debris // layer-by-layer construction // indentations possible only to a limited degree

Sustainability aspects uses waste materials // efficient use of materials for building components

GENERATIVE MANUFACTURING WITH RECYCLED MATERIALS

PROCESS PRINCIPLE

Vander Kooij presented a method for producing furniture similar to **fused deposition modeling,** using shredded synthetic waste taken from **old fridges,** and a robotic arm. The synthetic particles are fused in a receptacle to form a viscous mass, after which the **robot arm** traces the outline of a layer of the component, applying the material in strips using a jet nozzle. The material eventually cools and hardens into the desired outline. Molded forms emerge as the layers bond together. The jet nozzle is able to smooth each layer into the desired thickness. However, this means that undercuts are only possible to a limited degree.

Rapid prototyping of furniture using a robotic arm (Source: Dirk Vander Kooij)

Since Vander Kooij's experiment, scientists have discovered how to generatively manufacture three-dimensional objects using waste wood fibers combined with binding agents. One technology, developed by Rael San Fratello Architects in California, uses wood and cellulose fibers, which are combined during the 3D printing process and transformed into molded parts.

APPLICATION

Dirk Vander Kooij designed an entire manufacturing process to create his own furniture collection. His method allowed him to easily tailor the product according to specific requests and differing color variations. These principles of operation can be used in other sectors to make synthetic parts.

Color variations in layered structure (Source: Dirk Vander Kooij)

Bevel Bowl 3D printed wooden bowl (Design: Rael San Fratello Architects)

FDM principle (Source: Handbuch für technisches Produktdesign, Heidelberg 2011)

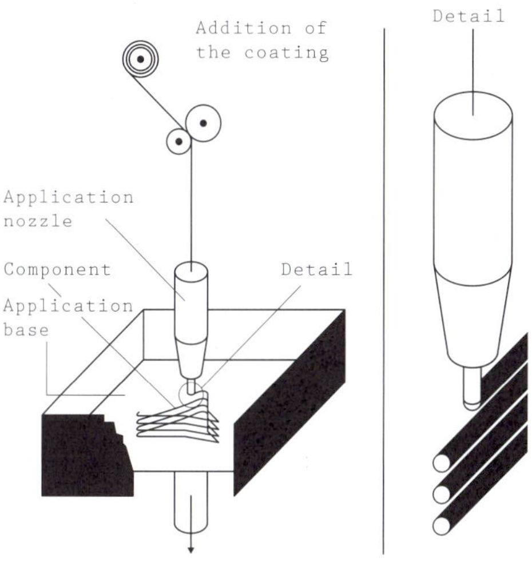

Fused Deposition Modeling

So-called rapid technologies were once developed to create prototypes of highly complex molds in the shortest possible time. Scientists nevertheless dreamed of one day being able to bring these technologies into the home for the manufacture of small objects for everyday use. Now it appears that research has brought this possibility a step closer to reality as, at the Vienna University of Technology, a 3D printer has been created in miniature. It is the size of a milk carton and will not cost more than €1,200.

Properties hardening of component parts in resin bath // layer-by-layer construction // use of light-emitting diodes // high-precision miniature components // possibility of manufacturing in nanoscale

Sustainability aspects efficient use of materials // possibility of creating biodegradable parts // energy-saving LEDs

3D PRINTING IN MINIATURE

PROCESS PRINCIPLE

The printer uses liquidized resin, which is selectively hardened by high-intensity light from **light-emitting diodes.** Layer by layer, the molded part emerges from the **resin bath.** The building platform is lowered after each individual layer is exposed to light. Each layer is only 0.5 mm thick. Alongside the traditional system which uses resin, researchers have also been successful in implementing the new printing techniques using biodegradable materials.

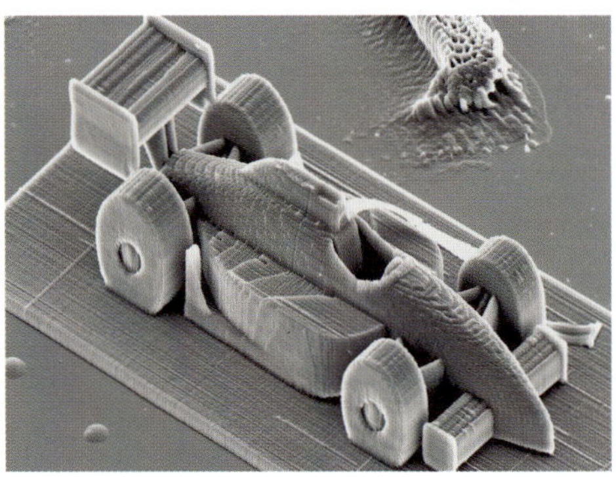

3D-printed structures in nanoscale (Source: Vienna University of Technology)

Miniature 3D printer (Source: Vienna University of Technology)

APPLICATION

The developers say that the system is able to produce even very small parts with high precision, making it particularly interesting for sectors that require highly complex and unique components on a very small scale, such as medical technology in the production of **hearing aids** and **skeletal components.** In addition, structures could also be produced for use within the body itself to stimulate bone growth. Scientists in Vienna attracted great interest in the spring of 2012 when they created objects in nanoscale using a 3D printer, greatly increasing the potential of **rapid technologies** for wider use.

Properties continuous manufacturing // layer-by-layer construction // unlimited length // 600 dpi resolution // high powder reuse rate

Sustainability aspects efficient use of materials and individualized production

CONTINUOUS 3D PRINTING

Whereas the possibilities for former generative manufacturing systems were limited by the amount of installation space available, since late 2012 the alternatives available in the field of mass production have no longer been subject to this limitation. It was at this time that the EuroMold 2012 was first introduced, a continuous 3D printer.

MATERIAL CONCEPT AND PROPERTIES

This leap in development was made possible by a new construction method that features a horizontal conveyor belt that coordinates how layers are constructed. At the beginning of the band, materials are applied layer by layer and, at the other end of the system, the finished part is removed. The system diagram shows almost unrestricted dimensions. The **maximum dimensions** are

850 × 500 mm. It is possible to achieve layers of up to 150 μm and 400 μm at a **resolution** of 600 dpi with a **high recycling ratio** for the powder used.

APPLICATION

This new system will help greatly to expand the potential of generative manufacturing for use in the mass manufacture of personalized products. The system is designed in such a way as to make construction and extraction possible in parallel without having to interrupt the process. Manufacturers say that this will help to reduce acquisition and operating costs compared with alternative systems. In spring 2013, designer François Brument collaborated with voxeljet to unveil a living-room concept that was entirely created using a 3D printer.

3D printer VXC800 in continuous operation (Source: voxeljet)

"Habitat imprimé" created by 3D printer (Design: François Brument, photo: Aurélien Dupuis)

Properties made from renewable raw materials // layer-by-layer construction via SLS (selective laser sintering) // elastic and flexible // water and alkaline-soluble

Sustainability aspects uses recycled materials // partly biodegradable

NEW MATERIALS FOR ADDITIVE MANUFACTURING TECHNOLOGIES

The development of generative manufacturing technologies – from a niche technology for building prototypes from complex molds into a technology for the mass production of individualized articles – has been aided by research into new materials. This work has greatly increased what is possible in terms of product manufacturing and also responds to the call for more sustainability in our culture of production.

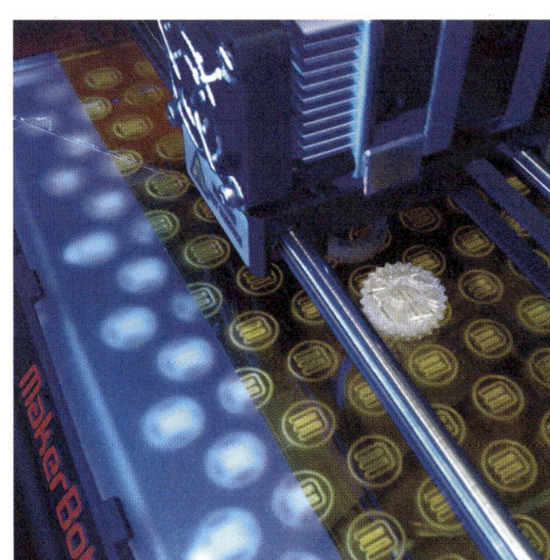

PLA components created by additive manufacturing methods (Source: MakerBot)

MATERIALS AND PROPERTIES

With its **Replicator™ 2,** MakerBot focuses on the use of biodegradable **bioplastic polylactic acid** (PLA) as the primary printing material. Acrylonitrile butadiene styrene (ABS), which is currently the most commonly used material, would only be used as a back-up material in the manufacture of molded parts.

At the end of 2012, a whole range of new synthetic materials for additive manufacturing was launched for use in **laser sintering.** Foremost in the developers' mind was ecological compatibility and resource efficiency. The new material PA 1101 is a naturally-colored **Polyamide 11** and is created from renewable raw materials. This material has a high impact strength and elongation at break and is particularly suited to operational elements such as integral hinges or components which are subject to impact stress.

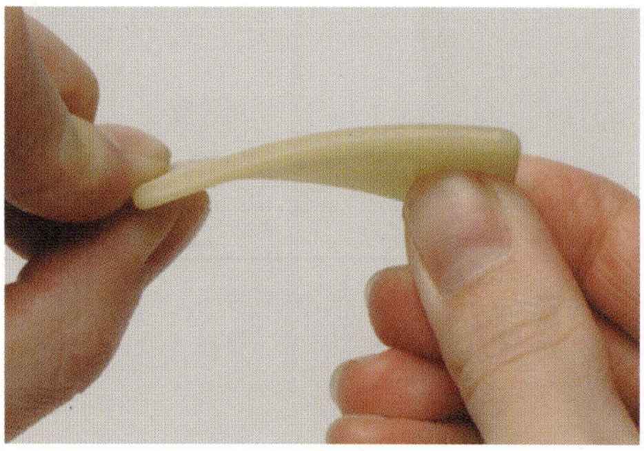

Flexible material made of PrimePart ST or laser sintering (Photo: Diana Drewes)

Munich-based EOS used mostly recycled materials to make PrimePart PLUS (PA 2221). The material requires only 30% virgin powder. This means that small amounts of leftover powder can be reused in the construction cycle, which leads to higher material efficiency in the laser sintering process.

Another material on offer from EOS gives product developers further possibilities, since it allows flexible molded parts to be created. PrimePart ST is a thermoplastic elastomer made from polyether block amide **(PEBA).** For the very first time, it gives designers the option to additively manufacture elastic connecting elements, seals, or buffers through laser sintering.

Belland is a new material for creating generatively manufactured interior structural supporting components which can be used instead of the usual choice, polyvinyl alcohol **(PVOH).** It can be made to dissolve in either water or alkaline solutions and is heat resistant up to 130 degrees Celsius.

Various materials for laser sintering (Source: EOS)

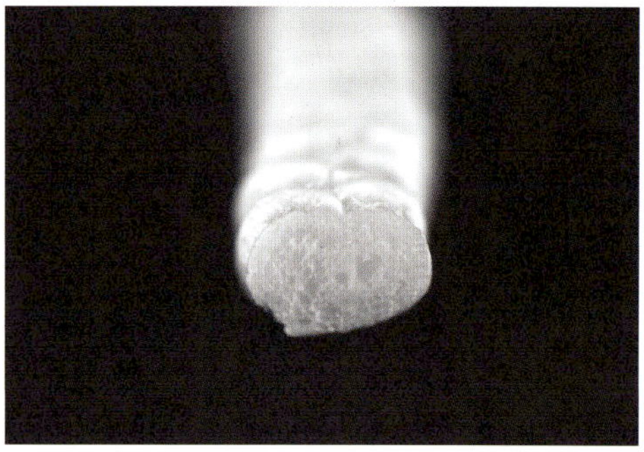

Alkaline-soluble foam (Source: BELLAND)

Properties layer-by-layer construction of organic tissue // meat replacement product // leather can be additively manufactured // skin transplants for treatment of burns

Sustainability aspects more environmentally friendly meat production

BIOPRINTING

Scientists and start-ups in the US have been making waves in the past few months with groundbreaking developments and are establishing a whole new context for the application of additive manufacturing. They are aiming to use 3D printers to create organic tissue.

DEVELOPMENT APPROACH

Meat fresh from the printer

American company Modern Meadow announced in October 2012 that they had manufactured a cube of meat made from living **animal muscle cells** using generative production technologies. The cube was created layer by layer using a 3D printer and agglutinated using **"bio-ink",** which contains a variety of cell types. The **piece of meat** reaches its final consistency in a **bioreactor.** The technology should turn out to be significantly more environmentally friendly than meat from **cattle farming** and help to reduce **factory**

farming. This process would also allow leather to be produced using additive manufacturing methods.

Printed tissue for medical purposes

To create human tissue, scientists at Organovo have developed a "bioprinter" that can be used to print three-dimensional cell forms of organic tissue. Their aim is to manufacture organs for research purposes and human tissue for transplants. In 2011, scientists at Cornell University were able to successfully manufacture **human tissue** and an artificial meniscus with a **digital fabricator.** Researchers at the Wake Forest University in Winston-Salem are also researching **bioprinting** with the aim of creating human skin for the treatment of **open burns.** The bioprinter is equipped with a **laser scanner,** which measures the wound and creates a matching piece of skin based upon the scan. Scientists working with James Yoo have already printed a large section of skin onto a pig.

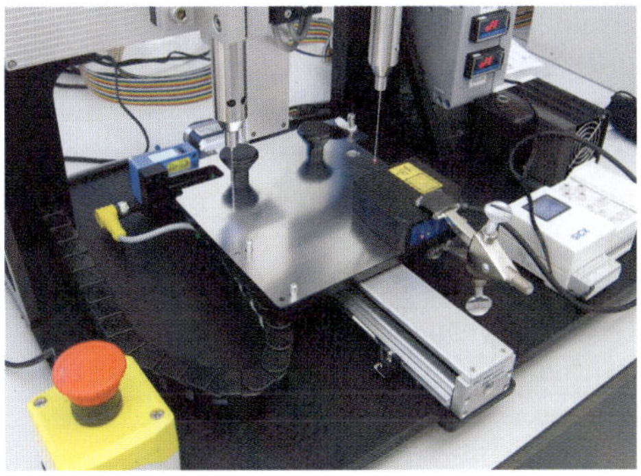

NovoGen bioprinter (Source: Organovo)

How bioprinting works (Source: Modern Meadow)

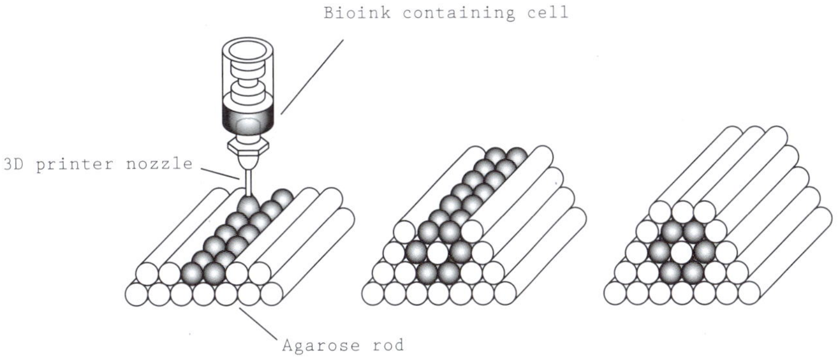

Properties durable markings // fusing of the surface of a plastic // alteration of the refraction index // possible incorporation of antibacterial and flame-retardant properties

Sustainability aspects makes printing processes more economical // structures created without the use of additional materials

In the past, lasers were science-fiction instruments in the imaginations of filmmakers and novelists. Today, however, the laser is a multifunctional tool that can be used to cut, bend, or weld materials. It's now adding another use to its repertoire: laser foaming, which can be used to make particularly durable markings.

LASER FOAMING

PROCESS PRINCIPLE

This technology is based on the fact that **laser radiation** of a plastic causes its surface to melt, resulting in **foaming.** Small **gas bubbles** arise that become fixed in the structure of the material, altering its **refractive properties.** Laser foaming has been optimized to such a degree that it is now possible to precisely create particular structures in a material's surface. Through the addition of certain additives, structures with antibacterial and flame-retardant properties can be produced. In addition, additives can be used to make products lighter and to reduce their environmental impact.

APPLICATION

Plastics with surfaces structured by laser foaming can be deployed in the creation of optical effects, signs, and lettering. Laser foaming can be used to create **Braille** characters, which could then be printed on the underside of door handles and accordingly guide blind people through a building. Laser foaming has already made inroads into the building industry, packaging materials production, and the automotive industry.

Wood is widely used in both interior and outdoor settings. However, because it is prone to the absorption and release of moisture, it is also susceptible to weathering. This occurs due to variations in the swelling of radial and axial grains. In cases where there is great variation, distortion of the material or fractures can occur. To prevent this, a number of tempering methods have been developed.

Properties limited moisture absorption // extremely great dimensional stability // dark coloration // resistant against fungus

Sustainability aspects tropical woods can be replaced by indigenous varieties // reduced energy required for procurement // greater longevity

WOOD TEMPERING BY WAX IMPREGNATION

PROCESS PRINCIPLE

The most widely used techniques are thermal. These temper the wood through a relatively long period of **heat treatment.** The resultant **"thermowood"** comes into being as the structure of the wood is altered by steam or submersion in hot oil. The cell structure of the wood changes as a result and its propensity to absorb moisture, as well as its propensity for swelling, is markedly reduced. Nevertheless, certain volatile compounds such as **resin** and **lignin** are lost during the process, which leads to the wood exhibiting a reduced dynamic load-bearing capacity.

With this new method, which uses a natural wax solution, these undesirable effects are negated. The technique is based on the discovery of Hamburg-born master carpenter Jan Nies, who in 2009 was able markedly to increase the **durability** of wood by treating its surface with beeswax. Following the heat treatment, natural **tree resin** and **wax** were pressed into the **wood cells** and partitions by means of a **pressure chamber.** These were then allowed to harden during a controlled cooling process. The treated wood has a low shrinkage and swelling rate, as well as great dimensional and formal stability.

APPLICATION

Tempering by means of **wax impregnation** is suitable for wood to be placed outdoors and to be used in street furnishings, playgrounds, and patios, and where it comes in contact with the ground.

Durable material – wood tempering using a natural wax solution (Source: Dauerholz)

Manufacturing three-dimensional objects using molding compounds, fibers, and particles is particularly interesting for designers. As themes of sustainability become ever more important, technologies have greater significance when they allow such molded parts to be manufactured in one go and also if they can be made using plant fibers.

Properties free, three-dimensional design // generative manufacturing process // possibility of working with waste fibers

Sustainability aspects conservation of resources // possibility of creating three-dimensional objects using natural fibers

THREE-DIMENSIONAL FIBROUS OBJECTS

Fabrican

Probably the easiest way to achieve this is to spray the fibers directly onto the object itself. British company, Fabrican, has developed a technique which makes this possible. The fibers, contained within a solution, are sprayed onto the desired object where they then harden into fibrous mesh. This method has gained a lot of attention, particularly in fashion circles. Yet there is enormous potential for use in interior design, medicine, and industry.

Organoid Design

This technology enables the manufacture of free-form objects using natural fibrous materials and organic binding agents. The customer can choose from a variety of materials such as **wood chips,** peanut shells, **pine needles,** or **scrap paper.** The construction process is such that it is suited to a variety of applications, from manufacturing small fruit bowls to spectacular office buildings. An **inflatable inner form** is used, which can be shaped as desired for the manufacture of the parts. The form is then sprayed with a porridge-like mixture of organic material and a natural binder. Finally the **biocomposite** is **vacuum** compressed and hardens into its final form. Organoid Technologies have already been able to create different types of seating and architectural structures from sustainable raw materials.

Building a bridge from sprayed sand (Source: Inder Shergill, Anna Kulik, Petr Novikov)

Kami Spin

To manufacture axially symmetrical molded parts from **paper pulp,** designers at ett la benn have created **rotation molds.** The initial mass made of cellulose fibers is first put into a **negative mold** and evenly distributed around as it rotates by **centrifugal force.** After drying, which does not require any additional energy sources, the molded parts are removed. The advantages of this method are that no unattractive seams are produced and that specific shapes can be transferred from the rotation mold onto the molded part in one step.

Concrete Canvas

Originally developed for military use, Concrete Canvas is a technology created from a **spacer fabric** filled with **dry concrete,** which can be flexibly formed and hardens completely within one day after contact with water. Its main uses are in the construction of stone buildings in crisis zones, and slope protection. The textile, available by the meter, can be placed over any surface and processed using the normal technologies. In this way, designers are able to realize furniture designs and architectural structures in stone extremely quickly.

Stone Spray

As part of the research project "Stone Spray," researchers are testing whether their method can be used to create highly durable structures from sand and a **binding agent** used in road construction. It is possible to create simple elements, but developers have focused particularly on directly manufacturing architectural structures.

Organoid carport (Source: Organoid Technologies, photo: Marion Luttenberger)

Kami Spin lampshade made using rotational molding (Source: ett la benn)

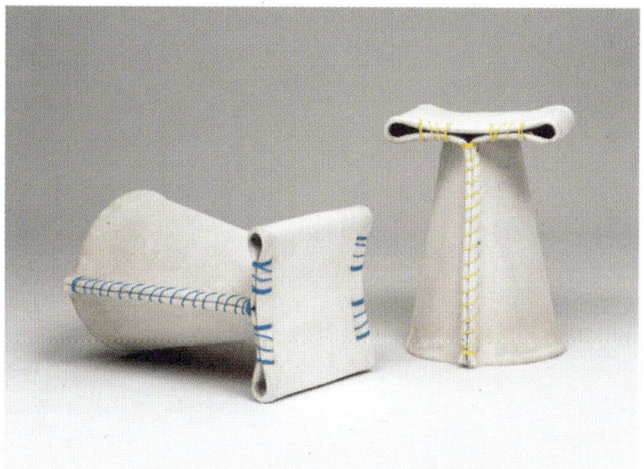

Stitching concrete – stools made using concrete cloth (Design: Florian Schmid)

BIOGENIC CERAMICS

Properties organic structures with the qualities of ceramics // manufacturing with liquid wood // transformation into green carbon-based materials // infiltration with silicon // similar hardness to a diamond

Sustainability aspects conservation of resources

One of the most interesting undertakings in the realm of bionic research deals with the transformation of natural, cellular, and porous materials into ceramic materials. The combination of a plant structure and the qualities of ceramic (extremely hard, high compressive strength, high temperature resistance) means it can be expected to have various applications. Researchers are working on a method which will allow the manufacture of high-performance ceramics in complex molded forms using wood-plastic composite (WPC) parts.

PROCESS PRINCIPLE

The potential of **biogenic ceramics** made from **liquid wood** components (30 – 70% wood) lies in their ease of manufacture through injection molding or extrusion techniques. In this process, the components are heated via pyrolysis to temperatures as high as 900 °C, during which they become **green carbon.** The original dimensions are mostly preserved and can even be further altered in the carbon state. Finally, the green matter is infiltrated with **silicon.** The original structure becomes completely saturated and the silicon bonds with the carbon within four hours to form **silicon carbide** (25% shrinkage). The hardness of the ceramic is almost equal to that of a diamond. The process can be compared to that of **wood petrification.**

Ceramic components made of WPC (Source: SKZ)

APPLICATION

Biogenic ceramic materials made of cellulosic raw materials are likely to be used in the future to allow cheaper manufacture of structural and functional ceramic products. Quite recently it became possible to manufacture ceramic pipes and contoured molded parts using the new techniques. Special WPC formulations with a wood percentage of 70% were developed for this purpose.

WOOD COATING

Properties powder-coating of wooden surfaces // scratch-proof finish // homo-geneous surfaces // possible to coat MDF

Sustainability aspects solvent-free coating // preservation of coating materials through single-layer application

Powder coating is a process most commonly used to make metallic objects scratch-proof. Until recently it was difficult to use this technology on wooden surfaces due to their natural properties (resin, shrinkage and sources below the fiber saturation point, ageing). Now, however, the process has been developed in Switzerland under the name "Woodcoat" to allow the coating of wood-based materials.

PROCESS PRINCIPLE

One of the greatest challenges in developing wood-coating methods was **electrostatics,** wood not being a conductive material. A **powder coating** is used that contains resins, pigments, and additives and which can be electrostatically charged and applied with a spray gun. Afterwards the powder-coated elements are run through an oven, where the powder liquidizes, the particles bond with one another and then harden. This technology demonstrates clear advantages in terms of environmental sustainability. This coating method is solvent-free, free of volatile organic compounds (VOC), and the superfluous powder can be reused. Furthermore, the method itself has a number of beneficial properties from a processing point of view. When applying traditional coatings, a number of layers usually have to be processed, since angles and edges in the wood tend to prevent easy application. These problems do not arise with powder coating, which results in even coverage.

APPLICATION

Currently, **medium-density fiberboards (MDF)** can be coated with a variety of powders. In the near future, developers are looking to extend the coating technology for use with further varieties of wooden materials.

Powder-coated wood-based material
(Source: Ramseier Woodcoat AG)

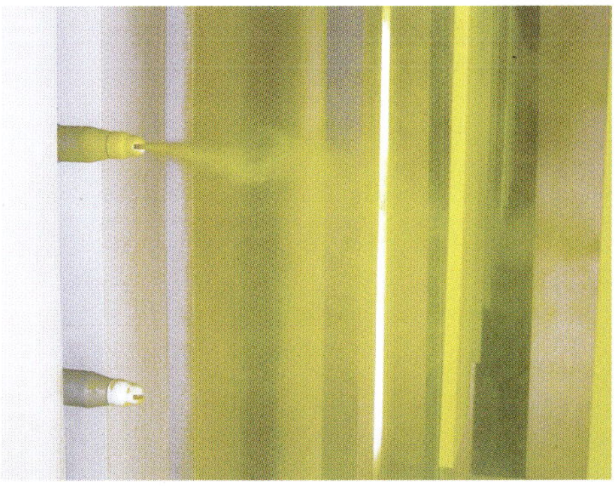

Process of powder coating wood
(Source: Ramseier Woodcoat AG)

Properties graphics for concrete surfaces // uses retardants // structures up to 1 mm thick // best results on pale cement

Sustainability aspects transferral of complex graphics with high material efficiency

GRAPHIC CONCRETE

In recent years, innovations in the construction sector have created ever more possibilities for modes of communication to be employed on traditional construction materials. Examples include light-transmitting and retro-reflective concrete, with which special optical effects can be achieved by integrating fiberglass in the solid mass or micro glass beads in the surface of the material. A further technology has been developed by Finnish manufacturer Graphic Concrete to allow the transfer of personally conceived graphics, images, or texts onto concrete surfaces.

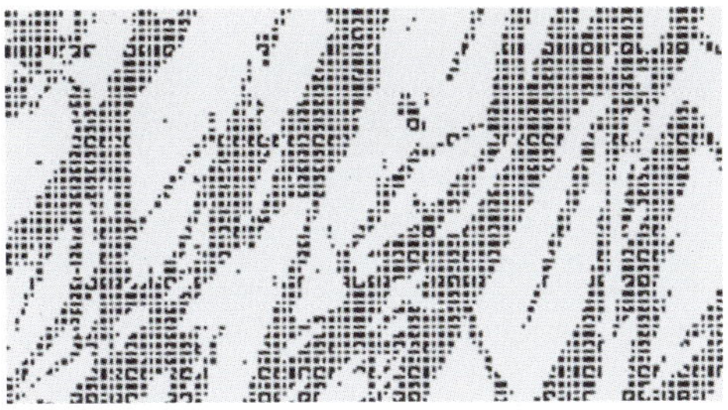

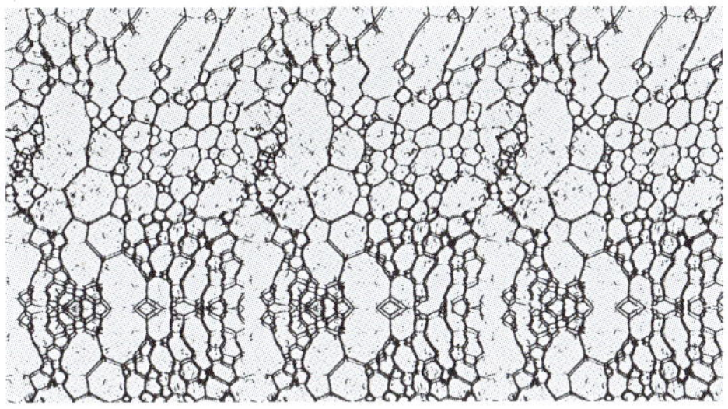

Potential pattern for the process
(Source: Graphic Concrete)

Interior use of graphic concrete
with plant-like structures
(Source: Graphic Concrete)

Hämeenlinna Provincial Archive has an
outer facade made of graphic concrete
(Source: Graphic Concrete)

PROCESS PRINCIPLE

The effect is created using a foil treated with a **retardant,** which is used to lay out the design of the components during the manufacturing process. The graphic elements and structures, which are up to 1 mm thick, emerge during the drying process due to the differences in contrast between the fine-grained **aggregate** and the foil. Previous applications showed that the best results were achieved on **white cement.** Should it be necessary to store the membrane foil, it should be kept in a dry and warm place.

APPLICATION

The process is suitable for use outdoors or indoors and has been used successfully on a number of buildings. Using this technique, local attractions and features can be transferred onto the surfaces of concrete buildings.

To achieve weight reduction in the field of aviation, metals are increasingly being replaced by artificial materials, and multi-material structures are being used. To easily create composite metal-and-thermoplastic structures, the GKSS Research Centre in Geesthacht has developed friction rivets, which allow joints to be created in a time- and resource-efficient manner.

Properties weight reduction // multi-material construction // rotating rivet penetrates the synthetic material // thermal insulation of the synthetic material // form-fit connections // lap joints and point-to-point connections

Sustainability aspects conservation of resources // no additional parts or adhesives required

FRICTION RIVETING

PROCESS PRINCIPLE

In friction riveting, a metal **rivet** is set into rotation and driven through a thermoplastic material into the surface of a polymer. The warmth generated by the tip of the rivet softens the thermoplastic material and due to high axial thrust allows the thermoplastic material to be penetrated. The contact force increases and the heat insulating properties of the thermoplastic material, which cause the tip of the pins to rise starkly in temperature, result in a deformation of the rivet. The rivet remains embedded in the plastic, producing a **form-fit connection** between the two materials.

APPLICATION

The technique can be applied to joints made of the technical **plastic polyetherimide** (PEI) and Al-Cu-Mg alloys. Friction riveting is particularly suited to the production of lap joints and point-to-point connections, replacing the need for traditional adhesive bonds.

Friction riveting in action
(Source: Sergio Amancio, Helmholtz-Zentrum Geesthacht)

How FricRiveting functions
(Source: Sergio Amancio, Helmholtz-Zentrum Geesthacht)

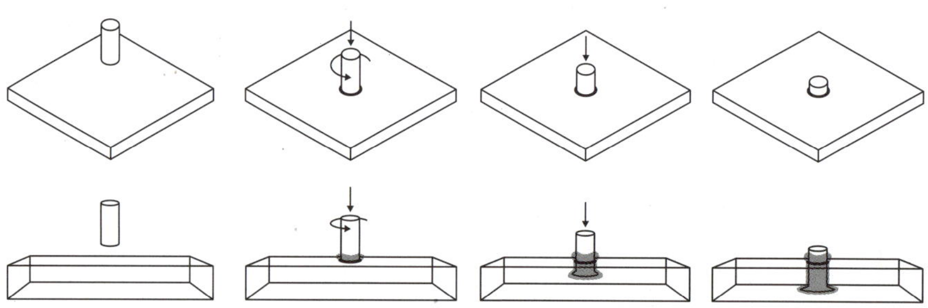

Properties processing of thin-film PV modules // nanoscale surfactants // breaking down of interfacial tension // separation of metal and glass // surfactant solution can be reused multiple times

Sustainability aspects recovery of rare and valuable resources

SURFACTANT-BASED SEPARATION PROCESSES

Today's common high-tech products demand more and more resources. For example, photovoltaic (PV) elements, battery systems, and even smartphones (which require up to 15 components using rare earths) contain a whole range of expensive metals, most of which must be obtained outside Europe. In this case, the problem is how to retain these for future usage and recycle them after the product's life span is over. The methods of recycling commonly in use today are often no longer suited to the vast raw material potential of waste flows. For example, PV cells are often processed with waste glass. Using this new technology, the constituent parts of high-tech products such as glass, plastics, metals, and semiconductor films can be separated from one another.

PROCESS PRINCIPLE

The new separation process from Saperatec in Bielefeld uses **nanoscale surfactants** to separate multi-layered compounds to retrieve valuable and rare materials. Surfactants might be familiar due to their use in household cleaning materials. As the **interfacial tension** is broken down, materials are released from a surface, after which they can easily be washed off. Saperatec GmbH was founded in 2010 and has applied this method to split up composite materials made up of glass, metal, and/or plastics. In this way, coatings and adhesives can be removed at room temperature. The **surfactant solution** can be used multiple times and is cleaned and circulated in a closed circuit.

APPLICATION

This method will first be used to process **thin-film PV modules** to separate **indium, germanium,** and **cadmium telluride** for reuse. The method is nevertheless also suitable for packaging materials and all manner of other composites, as well as for the retrieval of rare earth elements.

Separation of glass and metal using nanoscale surfactants (Source: Saperatec)

	Crystalline silicon	Amorphous silicon	CIS	CdTe
Glass	74	90	85	95
Aluminum	10	10	12	<0.01
Silicon	~3	<0.1		
Zinc	0.12	<0.1	0.12	0.01
Lead	<0.1	<0.1	<0.1	<0.01
Copper	0.6		0.85	1.0
Iridium			0.02	
Selenium			0.03	
Tellurium				0.07
Cadmium				0.07
Silver	<0.006			<0.001
Polymers	~6.5	10	6	3.5

Composition of PV cells: in percent (data from [BINE Informationsdienst 2010]: further details in [Sander et al., 2007]). An average weight can be assumed of 25 kg/m² or 15 to 20 kg per module

ADVERTISEMENTS

Evonik Industries

LIGHTWEIGHT DESIGN WITH THERMOPLASTIC COMPOSITES AND HYBRID STRUCTURES

The trend towards lightweight design has been going on for years. Two of the most interesting developments in this context are adhesion promoters for hybrid components and thermoplastic composites.

Matrices for composites have so far been mainly thermoset matrices used in established processes. Used with the same reinforcing fibers, thermoplastic matrices allow for significantly shorter cycle times in component production, can be stored indefinitely at room temperature, absorb less water, and are particularly suitable for medium- and large-scale production.

The polyether ether ketone VESTAKEEP® is suitable as a matrix for unidirectional fiber layouts or woven fabric reinforcement fibers, making it possible to produce fiber composite materials with a thermoplastic matrix.

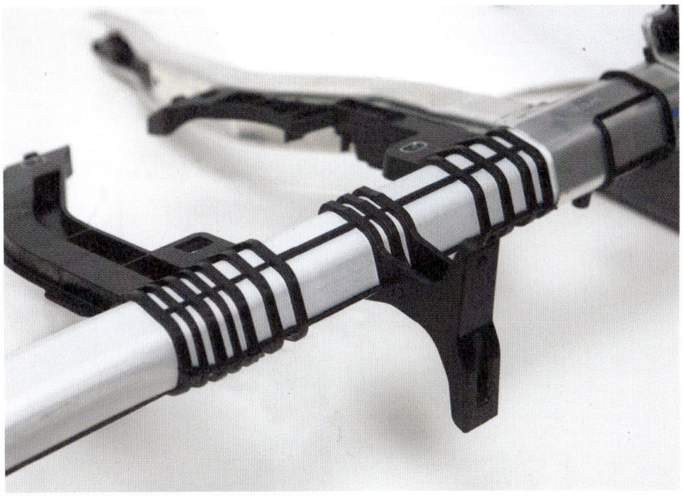

The polyphthalamide VESTAMID® HTplus has a very low viscosity, providing good fiber impregnation. With a glass transition temperature of 125 °C and a processing window of about 300 °C, this is a high-temperature matrix when combined with carbon, glass, or aramid fibers. If you are using a film-stacking process for manufacturing composite parts or sheets, suitable film of the requisite thickness is available.

VESTAMELT® is an adhesion promoter specially modified for metal-plastic hybrid parts. It significantly improves the mechanical properties compared with existing hybrid components, resulting in a weight reduction of up to 20 percent.

The adhesion promoter gives designers greater freedom, since they can use considerably less material when designing new components. Despite the substantial weight reduction, the components retain the required properties.

VESTAMELT® ensures a strong bond between metal and plastic and excellent sealing against the ingress of media.

EVONIK INDUSTRIES

Evonik Industries
Paul-Baumann-Straße 1
45772 Marl
Deutschland

Martin Risthaus
Phone +49 (0) 2365 / 49 43 56
Martin.risthaus@evonik.com
composites.evonik.com

High performance polymers for designers (iOS App)

Pfleiderer

A HEFTY ADVANTAGE : "BALANCEBOARD" IS MADE OF ANNUAL PLANTS

Materials made of wood are among the most sustainable there are, and include Pfleiderer's chipboards. Produced from the offcuts of sawn timber and wood removed in thinning, recyclable and healthy, they use the raw material wood optimally.

However, as demand for residual wood is steadily increasing, Pfleiderer has developed a new generation of wood-based panels: "BalanceBoard," based – naturally – exclusively on renewable raw materials. Just like tried and tested chipboard, it fulfils all requirements and properties based on DIN EN 312. The special feature of "BalanceBoard" is, however, that this new material is composed of around 35 percent of annual plants.

In the form of an innovative biomass granulate, it replaces around one third of the wood that would be necessary to produce conventional chipboards. "BalanceBoard" is therefore not only a wood-saving board, but is also far lighter than standard products; not only protecting the environment, but also facilitating further processing and use. "BalanceBoard" is particularly popular for interiors, for furniture and built-in elements, in shop fitout and trade fair stand construction as well as for hotel and ship interiors – as a raw board, direct faced ("DecoBoard Balance") or as HPL flat-bonded board ("Duropal Balance Flat-Bonded Element").

About the new Pfleiderer Group:
Pfleiderer is a leading European manufacturer of engineered wood materials. Group sales totalled more than EUR 1 billion in 2012. The total workforce of all group companies is around 3,500, with operations organized in two business units: the Business Centers Western and Eastern Europe. By combining the product ranges of Duropal, wodego and Thermopal under the umbrella brand Pfleiderer, the BC Western Europe is partner to industry, retailers, craftsmen, planners, and architects. The company has five production sites in Germany. The BC Eastern Europe includes the listed Polish subsidiary Pfleiderer Grajewo S.A., in which Pfleiderer holds a majority stake. Pfleiderer Grajewo S.A. has a strong position in the Polish engineered wood market.

PFLEIDERER

Pfleiderer Holzwerkstoffe GmbH
Ingolstädter Str. 51
92318 Neumarkt
Deutschland

Claus Seemann
Phone +49 (0) 91 81 / 28 630
Fax +49 (0) 91 81 / 28 252
claus.seemann@pfleiderer.com
www.pfleiderer.com

BARKTEX®

TEXTILES AND COMPOSITES FROM TREE BARK

Tree bark fleece is said to be the most ancient textile in the history of mankind. In 2008, UNESCO declared the postindustrial production process a World Cultural Heritage. Designers value the expressive character, unique texture and sensual tactility. The family venture BARK CLOTH® is pioneer of systematic bark cloth development and production and has been dedicated to the continuing cooperation with small-scale organic farmers since 1999. For its developments, it has been honored with a number of internationally recognized industrial awards for material engineering, design and social innovation. The permanently renewable bark of the East African fig tree is harvested every year without felling the tree. It is the base for a wide range of textiles and composites, which are manufactured in low-energy, partly CO_2-emission-free processes. They are distributed under the brand name of BARKTEX®. Use: interior finishes and durables.

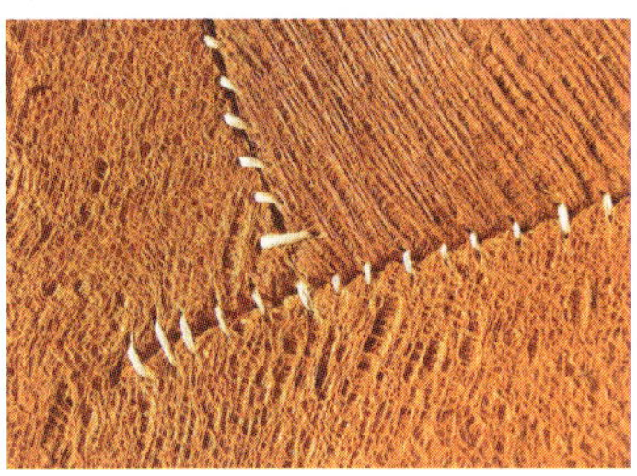

BARK CLOTH®_uganda Ltd.
c/o BARK CLOTH®_europe
Gewerbestr. 9
79285 Ebringen
Deutschland

Oliver Heintz, Mary Barongo
Phone +49 (0) 7664 / 403 15 60
Phone +49 (0) 700 BARKCLOTH
Fax +49 (0) 7664 / 403 15 61

barklcloth@barkcloth.de
www.barktex.com

HAVER & BOECKER

IMAGIC WEAVE® – TRANSPARENT MEDIA FACADES WITH ARCHITECTURAL WIRE MESH

IMAGIC WEAVE® transparent media facades offer the ability to create an interactive facade capable of effective visual communication, while at the same time serving as the building's second skin. IMAGIC WEAVE® combines Haver & Boecker architectural wire mesh with state-of-the-art LED technology and enables the creation of individually programmable lighting effects in all colors including full video presentation. The IMAGIC WEAVE® media screen appears as a homogeneous, elegant stainless-steel mesh facade, even when no content is being displayed.

The energy-efficient and environmentally friendly LED technology is just one contribution to ecologically sustainable building. Wire mesh facades ensure natural ventilation and daylight, which further reduces energy costs for air conditioning and electric lighting. Long life, easy and low maintenance ensure low running costs.

HAVER & BOECKER OHG
Architectural Wire Mesh
Ennigerloher Str. 64
59302 Oelde
Germany

Phone +49 (0) 25 22 / 30 684
Fax +49 (0) 25 22 / 30 767
architecture@weavingideas.com
www.weavingarchitecture.com

APPENDIX

Dr. Sascha Peters is the founder and owner of HAUTE INNOVATION – Material and Technology in Berlin. In the context of the services offered by his company he focuses on accelerating innovation processes and turning technological developments in materials into marketable products more quickly. Alongside leading technology companies such as BMW, Ottobock, Audi, and Evonik, his clients also include public institutions such as the Hessen Ministry of Economics, Technologiestiftung Berlin, and the European Commission. Since 1997 Dr. Sascha Peters has gained widespread expertise in product development, innovation management, construction, and industrial design. He headed research projects and product developments at the Fraunhofer Institute for Production Technology IPT, was deputy head of Design Zentrum Bremen, and head of the Modulor Material Competence Center in Berlin. In 2004 he was awarded a doctorate from the University of Duisburg-Essen for a thesis on improving communication between designers and engineers. Peters has authored numerous specialist publications. He lectures throughout Europe and runs workshops on innovative materials, sustainable materials, and energy technologies. In recent years he has held teaching positions at several German universities focusing on the subjects of material technologies, production, and construction.

www.haute-innovation.com

D

E

F

—

SELECTED PUBLICATIONS BY THE AUTHOR

—

11/2013
"Sustainable Multipurpose Materials in Design," in: "Materials Experience: Fundamentals of Materials and Design," ed. by Elvin Karana, Owain Pedgley & Valentina Rognoli, (imprint Butterworth-Heinemann, Elsevier).

10/2013
"Pappplattenpiloten – Design mit dem Akkuschrauber," in: Design Report, 5/2013, (Verlag Konrad Medien, Leinfelden-Echterdingen).

8/2013
"Design Fabriken – Designer gestalten Produktionsprozesse," in: Design Report, 4/2013, (Verlag Konrad Medien, Leinfelden-Echterdingen).

6/2013
"Reinbeißen statt Wegschmeißen – Essbare Verpackungen kommen in den Markt," in: Design Report, 3/2013, (Verlag Konrad Medien, Leinfelden-Echterdingen).

5/2013
"Inspired by Nature – Design Based on Organic Waste," in: GRID 4, (Institut für internationale Architektur-Dokumentation, Munich).

4/2013
"Metamaterialien," in: Design Report, 2/2013, (Verlag Konrad Medien, Leinfelden-Echterdingen).

3/2013
"New Timber Materials – Material Producers Respond to the Impending Shortage of Wood Procurement," in: GRID 3, (Institut für internationale Architektur-Dokumentation, Munich).

3/2013
"Gewachsene Möbel," in: md Magazin, 2/2013, (Verlag Konrad Medien, Leinfelden-Echterdingen).

2/2013
"Materialien für die generative Fertigung," in: Design Report, 1/2013, (Verlag Konrad Medien, Leinfelden-Echterdingen).

1/2013
"Changing Colors – Smart Colors for Designers," in: GRID 2, (Institut für internationale Architektur-Dokumentation, Munich).

12/2012
"Neue Kohlenstoffmaterialien," in: Design Report, 6/2012, (Verlag Konrad Medien, Leinfelden-Echterdingen).

11/2012
"Dancing for Energy – Energy Materials for Designers," in: GRID 1, (Institut für internationale Architektur-Dokumentation, Munich).

10/2012
"Hello Smart Materials," in: Design Report, 5/2012, (Verlag Konrad Medien, Leinfelden-Echterdingen).

9/2012
"Smart Energy Materials – Werkstoffinnovationen für die Energiewende," in: Schriftenreihe Nanotech, ed. by the Hessen Ministry of Economics, Transport, Urban and Regional Development, Wiesbaden.

8/2012
"Organische Gestaltung," in: Design Report, 4/2012, (Verlag Konrad Medien, Leinfelden-Echterdingen).

12/2011
"Handbuch für Technisches Produktdesign: Material und Fertigung – Entscheidungsgrundlagen für Designer," ed. by Andreas Kalweit, Christof Paul, Sascha Peters, Reiner Wallbaum, (Springer Verlag, Berlin).

11/2011
"Solid Lightweights – New Lightweight Solutions with Carbon Fibers and Natural Materials," in: form 241, (Birkhäuser Verlag, Basel).

11/2011
"Das Geheimnis des Betons," in: dds – Das Magazin für Möbel und Ausbau, (Verlag Konrad Medien, Leinfelden-Echterdingen).

11/2011
"Die Materialrevolution – Nachhaltige Materialien für Möbelbau und Interiordesign," in: MÖBELMARKT, jubilee issue, (Verlag Ritterhammer, Nuremberg).

10/2011
"Das zweite Leben des Gummistiefels," in: dds – Das Magazin für Möbel und Ausbau, (Verlag Konrad Medien, Leinfelden-Echterdingen).

9/2011
"Über den Tellerrand geschaut – Leichtbauwerkstoffe für Möbeldesign und Innenausbau," in: dds – Das Magazin für Möbel und Ausbau, (Verlag Konrad Medien, Leinfelden-Echterdingen).

8/2011
"Bambus und Banane – Nachhaltige Materialien für die Möbelbranche," in: dds – Das Magazin für Möbel und Ausbau, (Verlag Konrad Medien, Leinfelden-Echterdingen).

7/2011
"Biomimetic Material – New Materials Modeled on Nature," in: form 239, (Birkhäuser Verlag, Basel).

5/2011
"Extraordinary Timber – Wood Products with Innovative Qualities," in: form 238, (Birkhäuser Verlag, Basel).

5/2011
"Materialien einer neuen Designkultur," in: Design Anfang des 21. Jahrhunderts, ed. by Petra Eisele & Bernhard E. Bürdek, (Verlag avedition, Ludwigsburg).

4/2011
"Naturmaterialien – Veränderungen in der Werkstoffkultur," in: GIT Magazin, 4/2011, (GIT Verlag, Weinheim).

3/2011
"Natural Ingredients in Material Innovation," in: form 237, (Birkhäuser Verlag, Basel).

1/2011
"Materials Shape Products – Increase of Innovation and Market Opportunities with the Help of Creative Professionals," in: Schriftenreihe Nanotech, ed. by the Hessen Ministry of Economics, Transport, Urban and Regional Development, Wiesbaden.

SELECTED LECTURES BY THE AUTHOR

November 15, 2013
"Materials Revolution," ELMIA Subcontractor 2013, Jönköping, Sweden.

October 10, 2013
"The Materials Revolution: The Future of Construction," at the conference "The Stadium Design and Development Summit 2013," Nice, France.

October 4, 2013
"Smart City Materials," Smart City Expo BOGOTÁ 2013, Colombia.

September 17, 2013
"Materialinnovationen für Design und Produktentwicklung," VDID Industrietag, Composites Europa, Stuttgart.

June 8, 2013
"Smart Materials and Technologies," AEDES symposium "SMART City: The Next Generation," Berlin.

June 5, 2013
"Neue Geschäftsfelder durch Werkstoffinnovationen," for the network of the Bergische Entwicklungsagentur, Wuppertal.

May 28, 2013
"Living Light – Potenziale der Biolumineszenz für Designer," CTU, Prague.

May 14, 2013
"Smart Sustainable Materials – Werkstoffinnovationen für eine nachhaltige Produktkultur," Interzum 2013, Koelnmesse.

May 13, 2013
"Materialtrends für Designer 2013," VDID Industriedesign Tag 2013, Cologne.

April 22, 2013
"Smart Solutions for Advanced Living," Danish Design Centre, Copenhagen.

April 3, 2013
"New Material Technologies for Interior Design 2013," 3rd International Interior Design Conference, Moscow.

March 6, 2013
"The Materials Revolution," Ecobuild 2013, ExCeL, London.

January 31, 2013
"Light Materials – Materialien unterm Licht," Selux Licht Plus, Munich.

January 15, 2013
"Living Kitchen Materials," imm Cologne, Koelnmesse.

December 5, 2012
"Strategien für die umweltorientierte Materialauswahl in mittelständischen Unternehmen," specialist conference "Materialauswahl und Ressourceneffizienz," Hamburg Chamber of Commerce.

November 28, 2012
"Smart Energy Materials – Werkstoffinnovationen für die Energiewende," Euromold Werkstoffforum, Messe Frankfurt.

October 24, 2012
"Designing Energy – Smart Energy Materials," Coburg Connecting Conference 2012, Coburg.

August 23, 2012
"Biomimicry in Design and Architecture," Danish Design Center / Danmarks Tekniske Universitet, Copenhagen.

June 19, 2012
"Design im Wandel – Kreative als Impulsgeber für Technologieinnovationen," Standortagentur Tirol, Innsbruck.

June 14, 2012
"Sustainable Material Design for Automotive Interiors," automotive interiors EXPO 2012, Stuttgart.

May 31, 2012
"Surfaces for Future Mobility," BMW World, Munich.

May 16, 2012
"Nachhaltige Materialien und smarte Oberflächen für die Innenarchitektur," Conference on Interior Design, 2012, Haus der Architekten, Stuttgart.

May 10, 2012
"Material Revolution – Die Vision von einer nachhaltigen Produktkultur," New Design University, St. Pölten/Austria.

April 25, 2012
"Materialien im Kreislauf – Werkstoffe aus recycelten Resten und biobasierten Abfällen," Zukunfts-Agentur Brandenburg experts' forum, Hannover Messe.

March 21, 2012
"Materialien im Design 2012 – smart, leicht und unsichtbar," Museum August Kestner, Hanover.

March 14, 2012
"Leben 3.0 – Zukunft gestalten mit innovativen Materialien," "Leben 3.0 – Treffpunkt Zukunft," congress, CocoonClub, Frankfurt / Main.

February 22, 2012
"Werkstoffe für nachhaltiges Bauen," BAUTEC 2012, Berlin.

November 19, 2011
"Nachhaltige Materialien mit biochemischen Produktionsprozessen," sustainability trade fair, Postbahnhof Berlin.

October 12, 2011
"Naturnahe Materialinnovationen für eine neue Produktkultur," Alanus University, Alfter.

September 12, 2011
"Design drives Innovation – Die Bedeutung der Kreativindustrien für technologieintensive Innovationen," "create it 2011" conference organized by the Hessen Economics Ministry, Darmstadt.

July 4, 2011
"Materialien für eine nachhaltige Designkultur," Museum of Things – Werkbund Archive, Berlin.

May 26, 2011
"Material Revolution – Sustainable Multi-Purpose Materials for Design and Architecture," arena Designfestival, Poznan / Poland.

May 25, 2011
"Materials for a Weightless World," "Material Vision 2011" conference, German Design Council, Frankfurt / Main.

April 29, 2011
"Leichtbau Revolution – Eine neue Werkstoffkultur für Design und Architektur," Kunststoffland NRW, Düsseldorf.

April 4, 2011
"Biomimetische Materialien und Technologien für eine nachhaltige Zukunft," "GREEN-DESIGN 2.0" symposium, Haus der Kulturen der Welt, Berlin.

Translation from German into English: www.gainestranslations.de

Copy editing: Julia Dawson

Project management: Odine Oßwald

Layout, cover design, illustrations and typesetting:
Pixelgarten, Catrin Altenbrandt, Kai Bergmann, Timo Lenzen,
Adrian Nießler, Sebastian Pataki, www.pixelgarten.de

Printing: E&B engelhardt und bauer, Karlsruhe, www.ebdruck.de

Fonts: Neuzeit by Wilhelm Pischner, 1928 // Prestige Elite Std
by Clayton Smith, 1953 // Minion by Robert Slimbach, 1990
// Agency by Morris Fuller Benton, 1932

Paper: Cyclus Offset // Cocoon Offset // Woodstock

A CIP catalogue record for this book is available from the Library
of Congress, Washington D.C., USA.

Bibliographic information published by the German National
Library The German National Library lists this publication
in the Deutsche Nationalbibliografie; detailed bibliographic
data are available on the Internet at http://dnb.dnb.de.

This book is also available in a German language edition
(ISBN 978-3-03821-454-0).

P.O. Box 44, 4009 Basel, Switzerland
Part of De Gruyter

Printed on acid-free paper produced from chlorine-free pulp. TCF ∞

Printed in Germany

ISBN 978-3-03821-476-2

9 8 7 6 5 4 3 2 1

www.birkhauser.com

Print CO2 compensated
Ident-No. 106916